High Class Bar

출판사 : 모아미디어
저   자 : 지 애 린
발행일 : 2024. 08. 20
주   소 : 인천시 서구 건지로 28번길 107

High Class Bar

출판사 : 모아미디어
저   자 : 지 애 린
발행일 : 2024. 08. 20
주   소 : 인천시 서구 건지로 28번길 107
가   격 : ₩97,000원

Bar interior & Branding design?

When eating out, a good review isn't solely down to the food; a Bar interior plays an important role in the whole experience. Knowing how to create an environment that complements the menu and the space's interior architecture is no simple feat.

7 things to consider when designing a Bar.

1. Choose a Striking Colour Palette.

As with domestic projects, high up on the subject list for bar decor ideas is that of the colour scheme. "It's quite common for bar to be bold with their colour choices. In a sitting room or dining room there are just a few items of furniture and the colours chosen feel very much 'on show', in a bar setting, the volumes of furniture, accessories and light sources are far greater so there's a distraction from the colour. It's this that encourages the bravery and exploration in colour," Colour in a bar can also become a huge talking point. In larger establishments, it acts as a tool to define distinct areas. The colour becomes an icon for the bar in question. Remember, also, with colour decisions to reflect on how they sit together in daylight and how they evolve when night falls and they rely on candle and lamp light for luminosity.

2. Master a Functional Layout.

"A bar floor plan forms an immediate impression of ambiance. Keep tables spread out so that each feels secluded, and the guest experience will be an intimate one. "Tables packed tightly together on the other hand is a statement of conviviality and liveliness. Determining bar layout requires a decision on what atmosphere is hoped to be established,' There's a purely functional aspect to the bar room plan too though. The traffic should flow seamlessly. Bottlenecks are immediately apparent and cause an awkward distraction that guests will pick up on. A division between the bar and restaurant area is another question to be answered on the subject of layout. This will affect the feeling of formality or informality across the whole space.

3. Specify High-Grade Contract Furniture.

When a bar approaches its furniture choices as objects of function and nothing else, the entire experience becomes devalued. The importance of carefully selected bar chairs, tables and accent pieces is not to be underestimated. "It matters that not every element matches. That doesn't mean that every dining table and chair needs to be different, but that there are occasional pieces in the room to break up any consistency. A statement dresser or several elegant console tables adorned with decorative lamps or a vase are helpful here and serve as a reminder that bar furniture extends beyond table and chair."

4. Select Show-stopping Lighting.

Arguably one of the most crucial aspects of bar design, lighting ideas must be respectful of the fine line between necessary, task-style beams and ambient illumination. "In our homes, we consistently stress the need to layer the lighting throughout the space. It's no good having all of the lighting hung from above; it must drift down slowly from pendants to wall lights and lamps aplenty. A bar may be a commercial space, but guests want to be made to feel at ease in their surroundings as they do at home, so it's logical to follow the same lighting philosophy as you would in the home." Also acknowledges the effect that lighting has on each guest, considering what is the most flattering light at every angle. In the Harrods Dining Hall, they focus the lighting on the plate to showcase the food, with soft, low-level lighting for the diners to bask in.

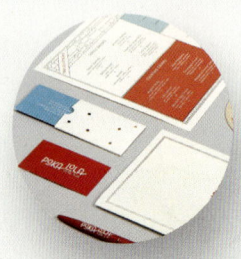

5. Curate a Unique Decor Collection.

"If there's exquisite detail in the food, then there should be exquisite detail in the way the room is put together." Restaurant decor ideas, therefore, are a fundamental component in how the approach every commercial project. "The term 'finishing touches' is misleading. These aren't the bits to be simply added in at the end. Accessories and decorative touches are where you deepen the level of consideration in bar interior design," Remember too that in the social media-centric world that we live in, carving out areas of appealing vignettes, complete with trinket boxes, decorative bowls and impressive floral arrangements, means your restaurant's design is all the more likely to become a must-visit and must-photograph location.

6. Display Personality-Full Artwork.

By extension of decor and accessories, displaying expertly curated arts and artefacts speaks volumes to guests. They serve as points of interest, they reinforce or subvert the overall interior design direction and they add warmth and texture to the space.
"Similar to how we light a bar with as much thought as we do a home's dining area, bar wall design deserves as much thought as would be given in a residential project. Art adorning the walls and sculptures atop of tables are luxurious details that reveal the pedigree of the establishment. We always encourage this in our bar interior design projects, and it's an investment that our clients never regret."

7. Don't Neglect The Bar Bathroom Design.

It's not uncommon to hear creative types and interiors aficionados claim that to know the true dedication to design of a bar, you must check out the bathroom. This is where the same level of attention to detail seen in the main bar either flails or flourishes. "The combination and contrast of materiality is so interesting in bathrooms. Marble and stone give a luxurious finish and contrast beautifully with timber, which can add warmth. Mirrored glass, brass and nickel all give reflection and a sense of luminosity." Lighting, too, is fundamental in this area of the bar, otherwise undoing all of the hard work in setting the perfect level of ambiance during dining.

Table of contents

008~019 Deep Bar YOD Group

020~027 Coquet fine living Cursor Design Studio

028~037 Ya Pan Bistro Baranowitz & Goldberg in collaboration + Pitsou Kedem Architects

038~053 Moon Club Formafatal + Machar&Teichman

054~063 Eduard's bar Stuttgart DIA - Dittel Architekten

064~071 Eduard's bar Düsseldorf DIA - Dittel Architekten

072~083 Beaubourg Brasserie Crown Creative

084~101 Punkraft bar Ater architects

102~127 Bar Marokana Loft buro

128~139 Rabbit Habit bar YOD Group

140~155 Orvay wine bar Isern Serra, Sylvain Carlet

156~163 Santo Cocktail bar IN-NOVA STUDIO

164~171 Mama Cocha Mario Sorrentino architect

172~187 Poka Lola Cocktail Bar CRÈME Jun Aizaki Architecture & Design

188~207 PAPA Dubai 4 Space Interior Design

208~217 VyTA Farnese Bar Collidanielarchitetto

218~231 Ostannya Barykada Lviv loft buro

232~245 Zweig bar Balbek bureau Architects

246~259 Merci Marcel Bar Hui Designs

260~271 ZWIN & SHOco YOD Group

272~287 More Fun Bar PIG Design (pigdesign.art)

288~299 Pearl, Chablis & Oysters BrandWorks

300~303 Bar St-Denis APPAREIL architecture

Deep Bar is a two-story venue in Dnipro city that unites a bar with a floristic shop and a tiny cafe. A small hall on the first floor has showcase windows that overlook one of the central streets. The flower shop with a few coffee tables locates there. In the big transparent fridge, you can see flowers side by side with wine bottles. The first floor works as the entering hall that leads to the body part of the venue in a roomy basement space.

Massive steel doors that look like the freezer doors hide the stairs that lead down to a bar. To be more specific, there are two bars inside Deep. The first one is a wine bar with a chamber-like atmosphere and a high vaulted ceiling. The second one is a mixology bar with lounge seats. Also, there is a smoking zone, a hookah zone, a partly opened kitchen, and a DJ place.

The emotional accent of the place is an installation made from timber slots that create a man's face. We placed it behind the bar counter, where usually you can see rows of liquor bottles. It is the metaphor of a peaceful person who is content with themselves and the world. Mostly people come to the bar to get that very feeling.

The main material in the interior is oak wood. We used some recycled material, boards that used to be applied in another building about 70 years ago. Honest touchable materials, along with the thoughtful light scheme, created a chamber-like cozy vibe. We turned the absence of windows in the space into an advantage. That is a perfect thing for such an evening bar format.

In the hall with the mixology bar, you can see some roots hanging from the ceiling. They conceptually connect two levels of the venue. They symbolize the roots of the flowers from the flower shop on the first floor. This way we reflect the idea of dipping that is connected with the name of the place. A flower can not be without roots, as well as anything in this world has its reasons and consequences. Everything is connected, and every action has some roots.

We started our work with an analysis of the advantages and disadvantages of the space. Based on its results, we offered our client the concept that we called 'Roots and Sprouts'. This idea shaped a legend of the place and emphasized the advantages of the space that we reflected in its interior. We highlighted its strong points of it and created a whole new conceptual format – cofounder of YOD Group Volodymyr Nepiyvoda says.

Project details

▶**Design:** YOD Group

▶**Homepage:** https://yoddesign.com.ua

▶**Area:** 437 m²

▶**Location:** Dnipro, Ukraine

▶**Photographs:** Andriy Bezuglov

▶**Branding and graphic design:** Pravda design

About us

Our story began in 2004, when studio's founder Volodymyr Nepyivoda and art director Dmytro Bonesko united their efforts and brought together talented architects and designers. Today we work predominantly in the sphere of commercial design and create projects in hospitality sector: hotels, restaurants, cafes and bars.

Taking on any new project, we want to become a reliable partner who is familiar with restaurant business and shares Client's vision. First of all, we search the idea which will be borne in people's mind and bring long-standing customers to you. To create a perfect image of the object, we design individually furniture, lighting and decoration, implement experimental solutions, create naming, graphic design and dishes serving.

Today we have the best restaurants in our portfolio and successful restaurateurs and hoteliers among our Clients. Our projects are published in famous magazines and win prestigious competitions. But the most important thing is that they bring our Clients love and loyalty of the guests.

The main material in the interior is oak wood.

We used some recycled material, boards that used to be applied in another building about 70 years ago.

Honest touchable materials, along with the thoughtful light scheme, created a chamber-like cozy vibe.

There are two bars inside Deep.
The first one is a wine bar with a chamber-like atmosphere and a high vaulted ceiling.
The second one is a mixology bar with lounge seats. Also, there is a smoking zone,
a hookah zone, a partly opened kitchen, and a DJ place.

013

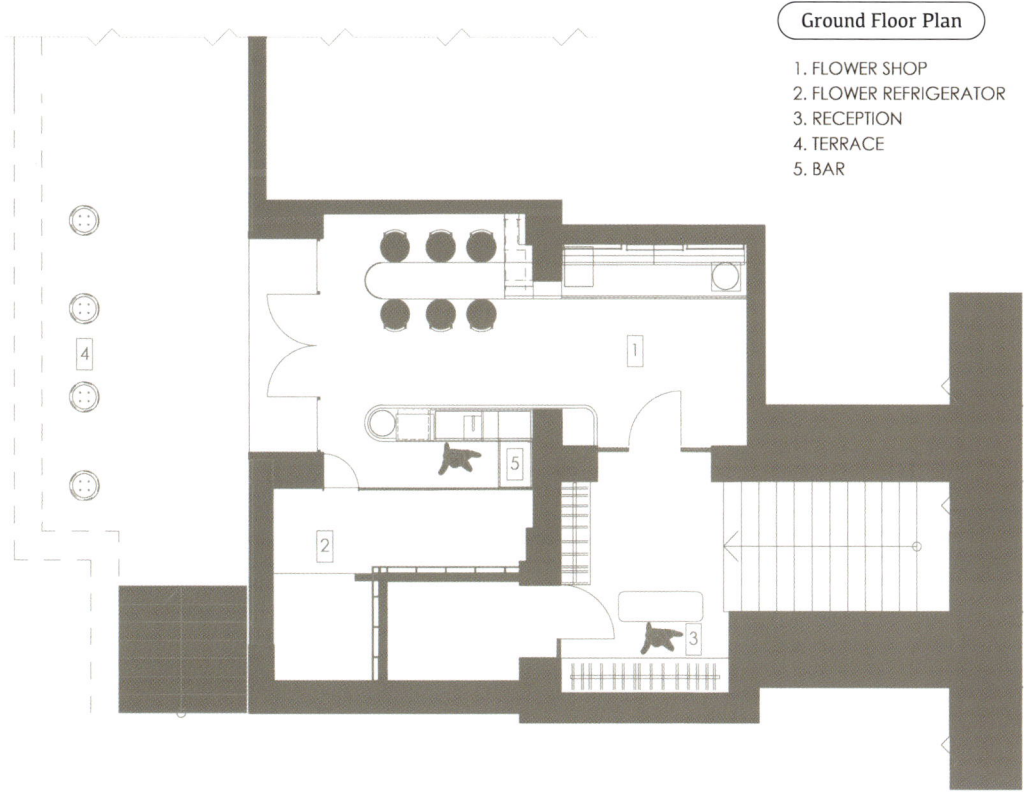

Ground Floor Plan

1. FLOWER SHOP
2. FLOWER REFRIGERATOR
3. RECEPTION
4. TERRACE
5. BAR

The emotional accent of the place is an installation made from timber slots that create a man's face.
We placed it behind the bar counter, where usually you can see rows of liquor bottles.
It is the metaphor of a peaceful person who is content with themselves and the world.
Mostly people come to the bar to get that very feeling.

First Floor Plan

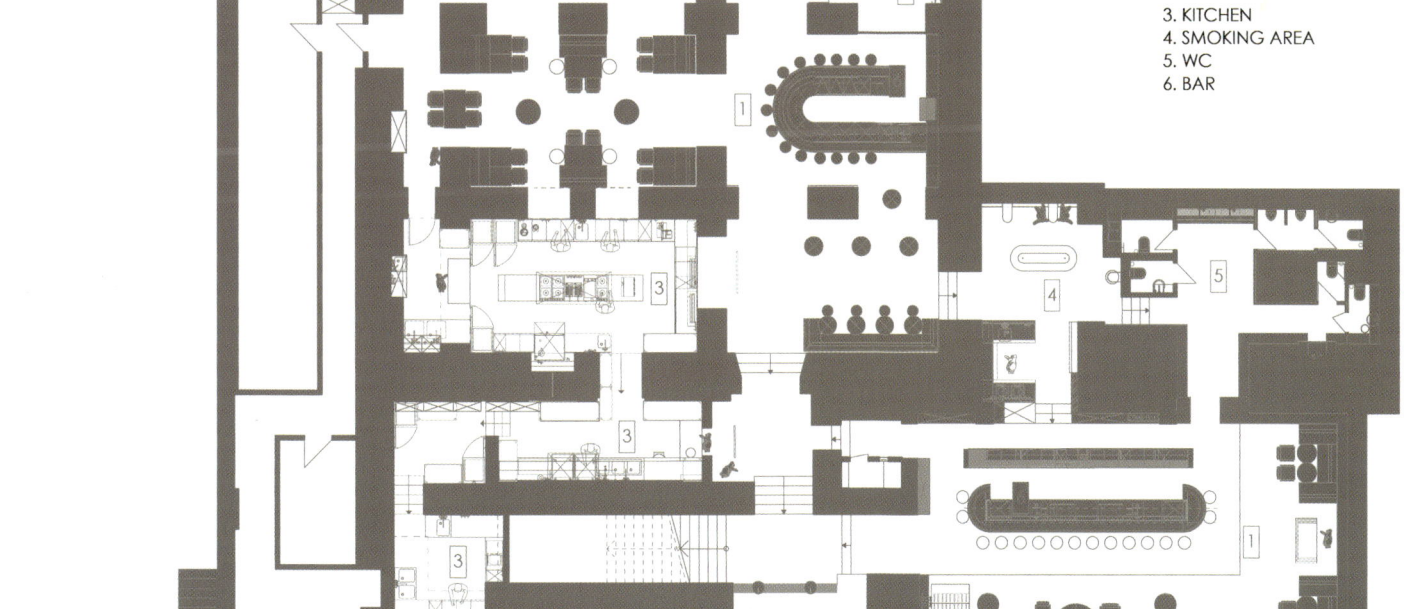

1. MAIN HALL
2. WINE REFRIGERATOR
3. KITCHEN
4. SMOKING AREA
5. WC
6. BAR

Larissa, Greece
Coquet fine living

For a downtown café-bar & restaurant: our new point of reference regarding quality entertainment. We installed a distinctive monogram at the stem of the logo, decorated with a red flower.

The passionate flower is rocking and embracing the monogram, until they become one elegant couple. Although embraced, our couple is breathing and maintaining the individual features unvaried. The logo is based on the top of the trademark, which is underlined by its promise: Fine living!

Project details

▶ **Design:** Cursor Design Studio
▶ **Homepage:** https://www.cursor.gr
▶ **Area:** 190 m²
▶ **Location:** Larissa, Greece
▶ **Photographs:** Michael Koronis
▶ **Interior design:** George Charakopoulos

About us

Successfully accomplishes the communication of your business, services and products to the domestic and global market. We consider each new mission as a challenge for our professionalism, talent, enthusiasm and passion: recognizable characteristics through out the full range of our work.

We face up branding as the ultimate communicative synesthesia and we faithfully serve all its functions: naming, logos, corporate identities, packaging, natural and digital promotion and art direction. The company founded in 2002 by Apostolos D. Tsiovaras.

▶ **Contacts** e-mail: info@cursor.gr / Call us: +30 2410 538458

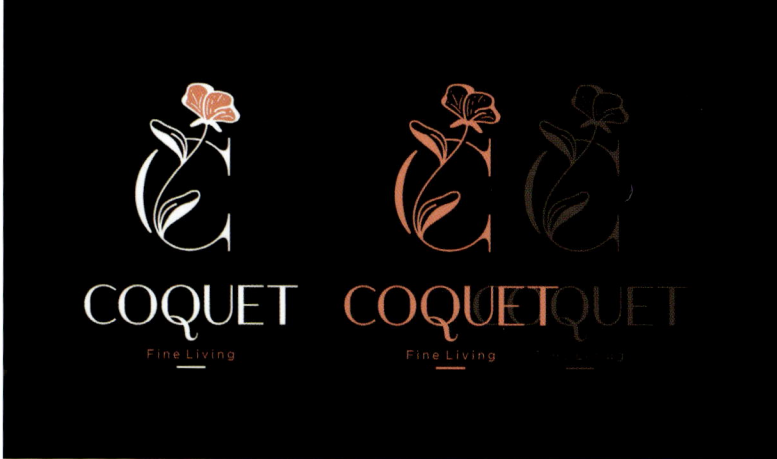

The passionate flower is rocking and embracing the monogram, until they become one elegant couple. Although embraced, our couple is breathing and maintaining the individual features unvaried.

Coquet fine living is an All day bar, offering fine living moments.
Coffee, inspiring food and coctails make your days and nights beautiful.

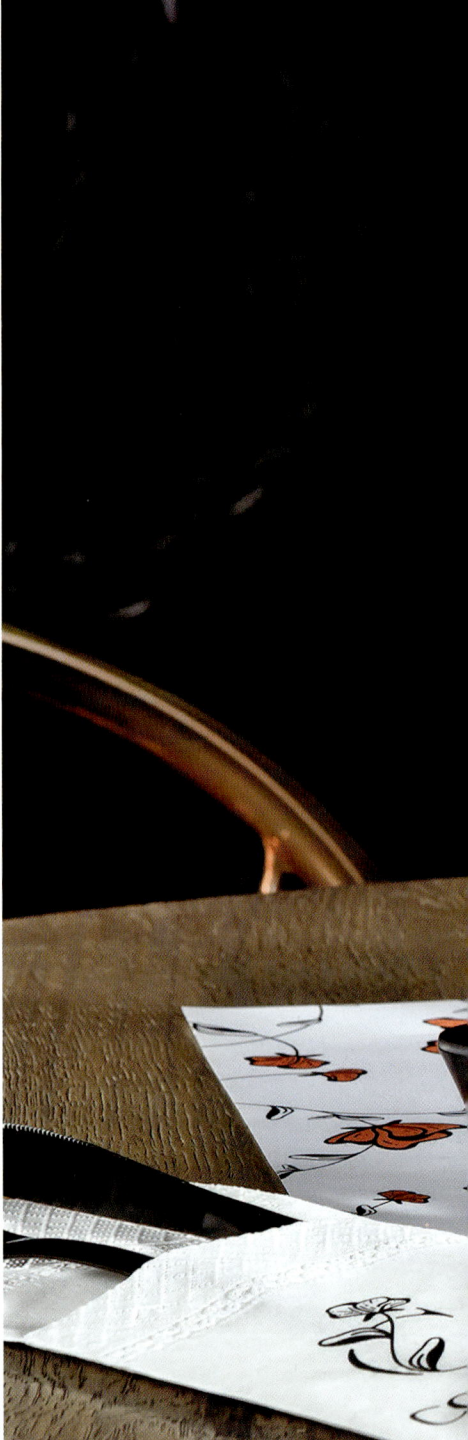

*A beautiful place with kind and helpful staff,
where you can enjoy your coffee, juice, lunch, lunch,
or drink in the evening. Many options in the catalog with relatively good prices.*

Ya Pan Bistro
Tel Aviv, Israel
Japanese Street Food Bar & Bistro

Ya Pan Bar & Bistro: Nachmani St 26, Tel Aviv-Yafo, Israel / +972 3-648-7796

YA PAN Japanese bistro is a new culinary hybrid that joins the Japanese izakaya with a western bistro. In both, people get together in an inviting and informal atmosphere to enjoy dishes from the traditional kitchen associated with their culture. The culinary experience at YA PAN is an expression of Chef Yuval Ben Neraya's personal interpretation to how these two worlds interact, and the design is yet another reflection of this marriage.

The desire to create an informal atmosphere together with the space's physical trait being long narrow and tall, resulted in a layout of one sweeping stroke in the form of a central bar. The bar dominates the entire space and facilitates dynamic social interactions.

Japan's kite festivals are a place for exciting social gatherings where groups of people get together to fly huge kites. This ancient tradition with its colorful aesthetics was the inspiration for the design of the space's vertical dimension. Colorful lit metal meshes hang randomly above the bar. Those at the bar enjoy different views around the bar that change depending on where one is seated.

Small, round mirrors are spread along all the enveloping walls around the bar. Their steady rhythm creates a visually pleasant pattern while the mirrors reflect a constantly changing reality depending on the location of the viewer. The mirrors is a gesture that recalls the mirrors used in western bistros, where they allow guests to feel part of overall experience in the space.

Project details

▶**Design:** Baranowitz & Goldberg in collaboration + Pitsou Kedem Architects
▶**Homepage:** https://www.baranowitz-goldberg.com
▶**Area:** 70 square meters
▶**Location:** Nachmani Street, Tel Aviv, Israel
▶**Photographs:** Amit Geron
▶**Lighting design:** Orly Avron Alkabes

About us

Having opened in 2017, Baranowitz & Goldberg partners Sigal and Irene have been collaborating as colleagues in many different commercial design projects. After a few years of successful joint work, the two architects established their own practice, paving a new way based around their diverse expertise.

Irene & Sigal's comprehensive experience in different fields of design allow the studio's work to vary and include all scales. Whether it be a product or a building, creating a story that will ignite the creative process is the leading principle of their work. The studio's projects are characterized with a playful balance between art and design while adhering to strict yet poetic principles of proportion, and a true integration between conceptual thought and functionalism. When created within a strong conceptual framework, creativity can turn to unpredictable yet coherent directions. The studio's goal is to allow each project to pave its way toward a unique architectural interpretation.

The large central bar dictates circulation through the space and encourages casual social interactions among the clientele. This also allows most of the indoor guests to be oriented towards the window, connecting the interior space with the exterior. Baranowitz explains, 'the street on which the Ya Pan is located is one of Tel Aviv's most lovely places. It was clear to us that we wanted to accentuate the connection between the outdoor scene and the interior.'

Making the most of the long and tall physical dimensions of the space, the large, colourful sheets of metal mesh that hang from the ceiling in Ya Pan make a visually striking statement, even through the windows. Inspired by Japanese kite festivals, the transparency of these elements combined with the strategic lighting creates a particularly ethereal effect at night, which is when bars and izakayas come alive.

YA PAN Japanese bistro is a new culinary hybrid that joins the Japanese izakaya with a western bistro. In both, people get together in an inviting and informal atmosphere to enjoy dishes from the traditional kitchen associated with their culture.

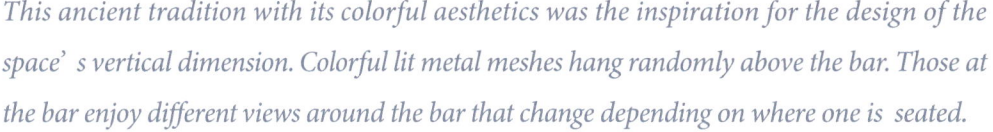

This ancient tradition with its colorful aesthetics was the inspiration for the design of the space's vertical dimension. Colorful lit metal meshes hang randomly above the bar. Those at the bar enjoy different views around the bar that change depending on where one is seated.

Repeating circular mirrors line the bare concrete walls, a detail that references a traditional bistro. These reflective elements enhance the ambiance and create a youthful, informal space, glittering with the colours illuminated from above.

Repeating circular mirrors line the bare concrete walls, a detail that references a traditional bistro. These reflective elements enhance the ambiance and create a youthful, informal space, glittering with the colours illuminated from above.

Moon Club: Dlouhá 709/26, 110 00 Staré Město, Czech Republic / +420 703 140 640

Moon Club
Prague, Czech Republic

The creative union of studios Formafatal and Machar&Teichman was given a chance to design the mysterious Moon club, located in a vibrant Prague city area of the Dlouhá street. The newly opened club brings a brand-new and original view of the interior design in this city area. Clients' vision was to create an interior with details that would be on a significantly higher level compared to other enterprises on the same wave.

The space of the club, which used to serve as a bank administration in the past, connects the central hall with a glass roofing. The element of the inner yard was one of the challenges of the spatial solution as well as the technical aspect of the acoustics which was given great attention. Acoustic measures, such as doubled roof glazing, acoustic buffer wall constructions and special plasters are incorporated into the interior concept. Different zones were created in order to overlap the original building layout with the needs of the club operation. Cosy seating areas with comfortable lounge furniture and mild light intensity are placed under arch ceiling alongside the central hall, while the dance floor is placed opposite to it. One has an overview of the happenings around the central bar from the courtyard gallery, can move into one of the lounges or enjoy the atmosphere in the alchemist bar located on the second floor.

Project details

▶**Design:** Formafatal + Machar&Teichman
▶**Homepage:** https://www.formafatal.cz
▶**Area:** 740 m²
▶**Location:** Prague, Czech Republic
▶**Photographs:** Jakub Skokan, Martin Tůma / BoysPlayNice

About us

We're team of friends – architects, designers and scenographers. We're creative studio focused on architecture, interior design, exhibition installation and product design. Team of ten professionals is currently working on several commercial and residential projects not only in Czech Republic, but also across the world. Formafatal studio has already won several domestic and international awards for their completed interior projects.

"We create public spaces, where people feel cozy, and homes, that are tailored to the clients needs. All projects we approach individually and with focus on specific human needs and client's requirements. Individual approach for each project is based on mutual understanding with the client, enthusiasm, natural collaboration and unified conceptual solutions. We solve projects complexly from creative concept to realization, with attention to detail. To us, our work means life and passion. In our projects, we often cooperate with other artists because we love to support talented individuals - together with them we create original and innovative spaces that cultivate society.

The main bar is covered with goldish corrugated metal cladding. The DJ overlooks the club from the back of the bar – a rounded wall made of burned wood. The club has a welcoming view from the street – looking through the entrance passage you can directly see the central yard with the bar. The first smaller bar placed in the entrance area welcomes the guests. The floor is made from wooden cubes as a reminder of historical function of the passage.

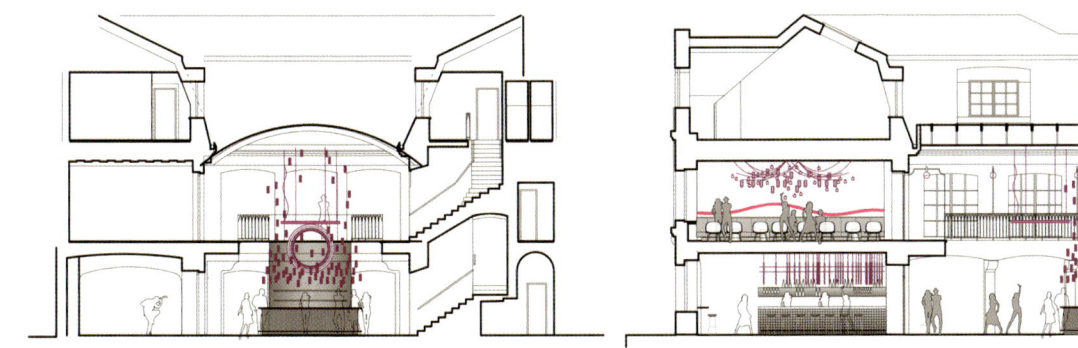

Main Bar Plan

041

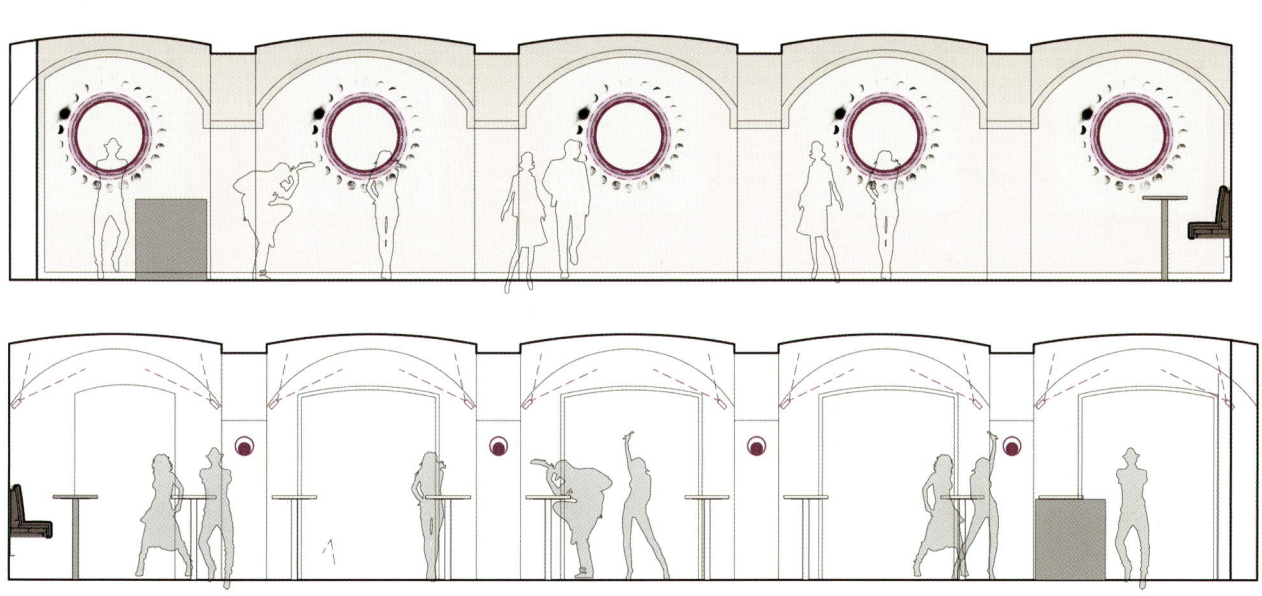

Elevation

043

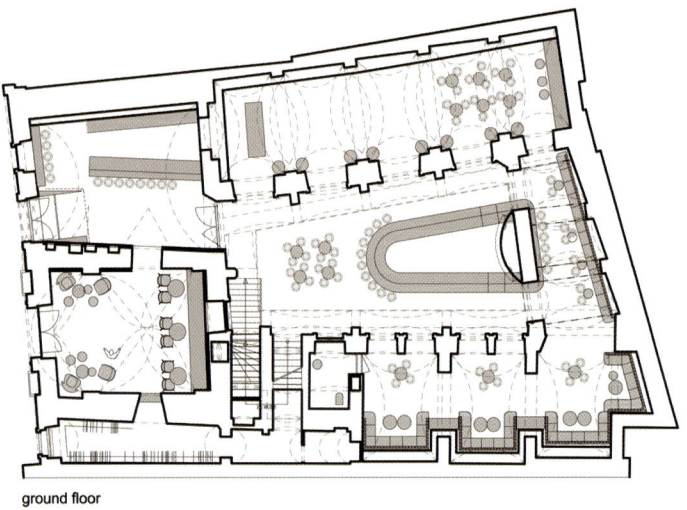
ground floor

1st floor

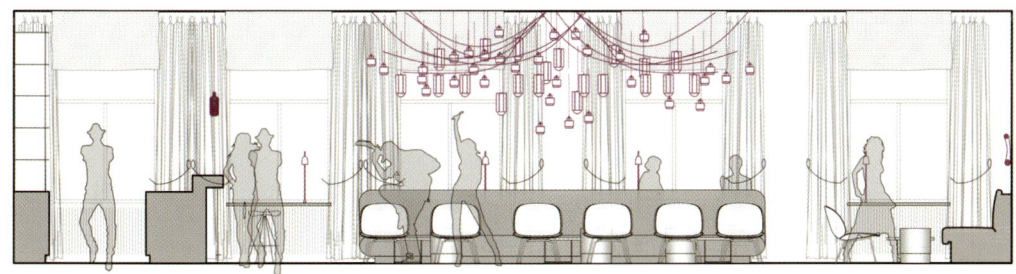

Elevation-b

Elevation-c

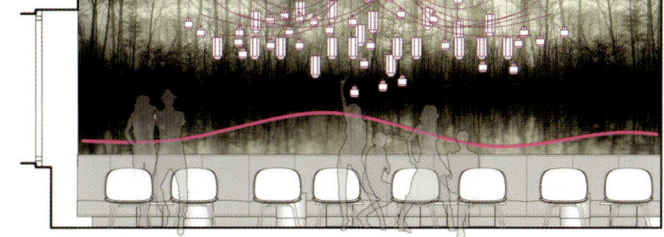

Elevation-d

047

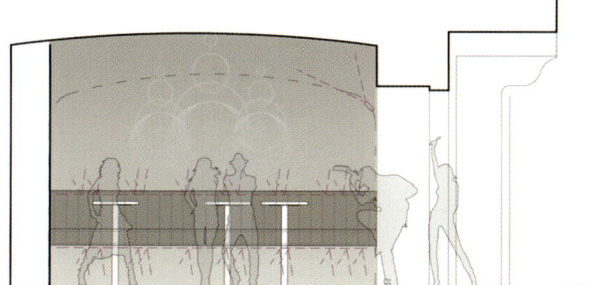
Elevation-f

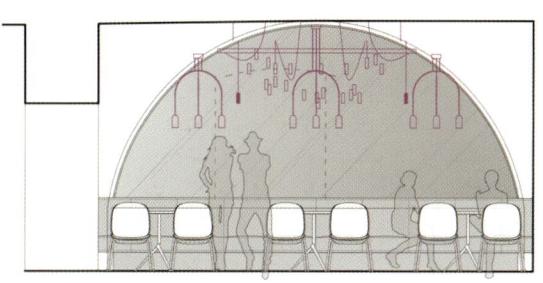
Elevation-g

Elevation-e

049

The name of the club is linked to the main motive – the Moon. Major topics are the mysteriousness of the nightfall and the alchemistic mystique. The walls of the 740m2 club are covered with patina-painting, which is a common link to all the spaces that vary in atmospheres and details. Materials used for interior's elements also vary through the rooms – patinated metal sheets, dark burned wood, old stained mirrors, ornament paintings on the walls and velvet upholstery furniture in several elegant colours.

Materials used for interior's elements also vary through the rooms – patinated metal sheets, dark burned wood, old stained mirrors, ornament paintings on the walls and velvet upholstery furniture in several elegant colours.

The club has a welcoming view from the street – looking through the entrance passage you can directly see the central yard with the bar.

Eduard's bar
Stuttgart, Germany
Lifestyle bar with a premium view

EDUARD'S

DIA – Dittel Architekten is responsible for developing the restaurant's design concept, the name Eduard's, its corporate design, and also for the construction of the bar in Stuttgart's Dorotheen Quarter. Named after the founder of the Breuninger department store, this new lifestyle bar is making a bold statement in the new quarter. The striking design provides a fresh perspective on a piece of history, making the counters the focal point of where it's at, much like the early days of bar culture.

Even from the outside, the bar is unrecognizable as such: as bold and elegant as a tumbler of Bulleit Bourbon, the modern, solid wood bar is located right at the entrance to the space. Once inside, patrons immediately find themselves within the bartender's field of vision, their first glass in hand.

The narrow, five-meter high space is a defining feature at Eduard's. Surrounded by a glass façade opposite a brick façade, the bar is divided into a brighter area in the front and a darker area at the back – designed to be welcoming by day and night. By day, spots along the glass façade are hugely popular, thanks to the natural light and a first-class view. When it gets dark outside, the bar becomes the focal point.

Project details

▶ **Design:** DIA - Dittel Architekten
▶ **Homepage:** https://di-a.de
▶ **Area:** inside area of 100 m2 and another 70 m2 outside
▶ **Location:** Stuttgart, Germany
▶ **Photographs:** Martin Baitinger

About us

In our architectural office, architects, interior architects and communication designers work together to make unique creations a reality. For every project, we go through an exciting creation process that is subject to our quality standards and that, in symbiosis with inspiration, well-grounded knowledge and teamwork based on trust, results in successful project implementation.

Many years of theory and practice have made us experts in every single phase of an architectural project. In conjunction with our digital knowledge and deep understanding of our industry, we are able to provide our clients with holistic advice and design. All this makes our architectural office a place where expertise and diverse experience of our team members come together. Great teamwork is characterised by respect, transparency and understanding. It forms the basis of promising work and excellent results. This applies to working with our clients and partners as well as to cooperative work within our architectural office.

When night falls, the defining copper tones, highlighted by shades of patina and the reflection of lights, create a one-of-kind ambience that glows invitingly when seen from outside. The bar becomes a welcoming, stand-alone feature. The rust-brown shelving accentuates the luminescent bottles of spirits and the hand-crafted iron lamps in the suspended ceiling become stylish overhead lights. The striking ceiling installation, consisting of delicate, angular rods and eye-catching bulbs, completes the picture. With a compact. inside area of 100 m2 and another 70 m2 outside, Eduard's offers seating arrangements to suit everyone' needs: traditional raised stools at the bar or at the convivi side tables, which diffuse the original purpose of the bar a barrier and allow patrons to be seated alongside th barkeeper. Those wishing longer conversations can either s in the comfortable vintage armchairs or gather in groups o the sofas and chairs, all of which look out onto the outdoc patio. The benches, conveniently surrounded by acousti panels, become casual lounge areas that overlook the ba The perfect spot to sit if you want to see what's going on.

e elegant character of the metal serves as a contrast to the
w industrial floor and the exposed concrete walls. Wall
sign elements and textiles in muted green tones, dark wood
d rustic leather upholstery complement the colour and
terial concept. The corporate design infuses every detail.
e name and logo unite tradition with the modern age and
e indicative of the timelessness of elegance, which is real-
d across the different eras. The geometric design reflects
e structure of the ceiling installation and is emphasized by a
ld and streamlined typeface: Welcome to Eduard's!

Floor Plan

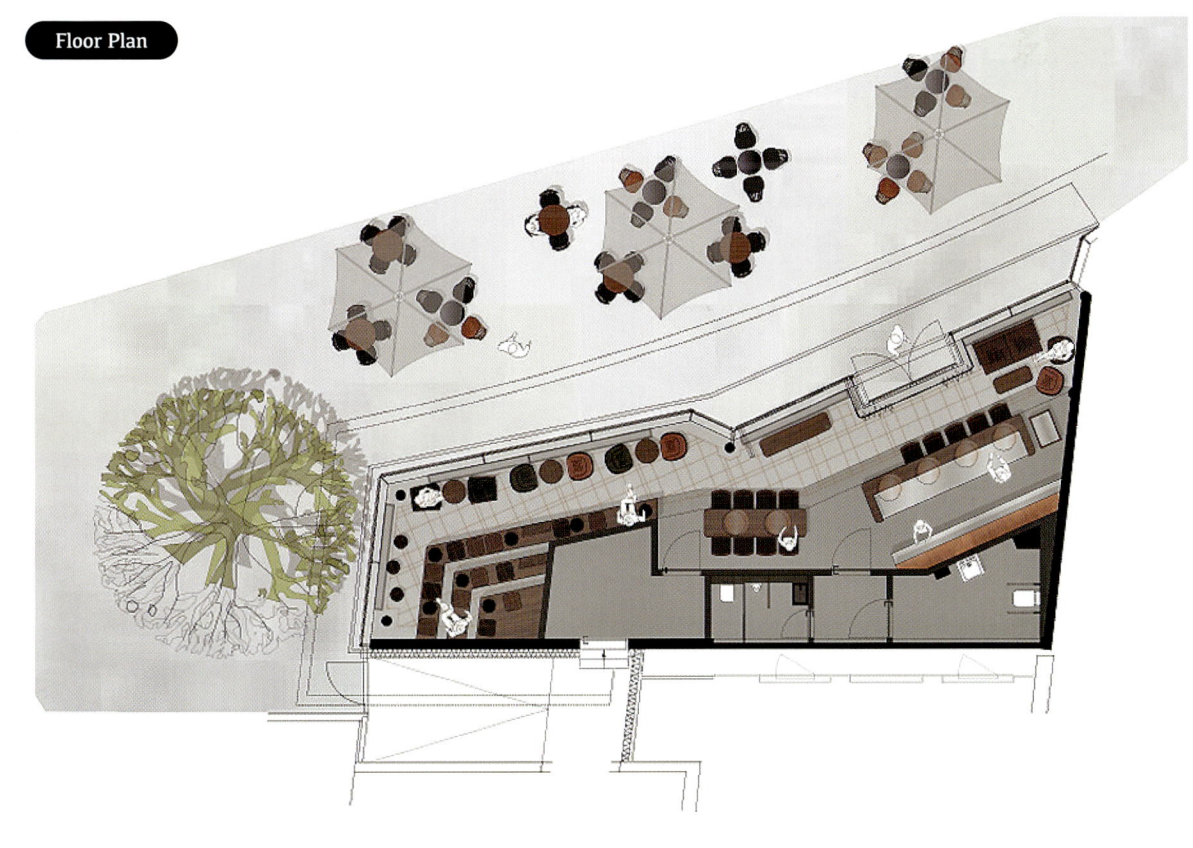

Wall design elements and textiles in muted green tones, dark wood and rustic leather upholstery complement the colour and material concept.

The striking ceiling installation, consisting of delicate, angular rods and eye-catching bulbs, completes the picture.

Eduard's bar: Kö-Bogen, Königsallee 2, Düsseldorf, Germany / +49 211 566414660

Düsseldorf, Germany
Eduard's bar
A modern spatial concept in a prime location

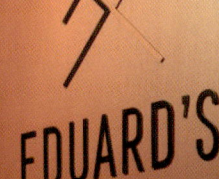

EDUARD'S

DIA – Dittel Architekten brings the concept of Eduard's Bar from Stuttgart to the Rhine. The chic new bar just opened in the heart of Düsseldorf's Old Town and is sure to impress guests with its creative menu and stylish ambience. DIA - Dittel Architekten was commissioned by the fashion and lifestyle company E. Breuninger with designing and implementing the project in the famous Kö-Bogen. The two companies already partnered back in 2014 when DIA realized the successful concept for Breuninger's Sansibar in the building ensemble.

Kö-Bogen, the futuristic structure of star architect Daniel Libeskind, is considered one of the architectural highlights of the city and provides a spectacular backdrop against which the bar can unfold its unique architectural potential. The tapered shape of the Kö-Bogen, with its fascinatingly dynamic structure, divides the external terrace into two sections, providing a thrilling view for guests and enticing passers-by to stop over for a while.

Eduard's Bar in Stuttgart has been immensely successful ever since its launch in 2018, and the design concept for it even received the prestigious Red Dot Design Award.

During the implementation phase, DIA utilized the high recognition value of this concept, but also referenced the vibrant art scene of the city and gave the bar a distinctive character by introducing specially selected graphics and art.

When visitors glance into the entrance area, their attention is immediately directed towards the interior by the appealing ceiling installation, graphics and lighting. The bar's harmonious color scheme – dominated by modern coppery overtones – and its combination with the exterior glass façade create a fascinating play of light and shadow. The reflection of the warm red colors in the windows of the bar adds an inviting ambience to the place. This warmth of the color scheme permeates the entire interior. The hand-crafted wrought-iron lamps with a golden finish, the ceiling installation made of rust-brown metal structures and the high-quality seating with a vintage look are all essential design elements with a high recognition value that have already proven their worth in the bar's Stuttgart counterpart: a timeless design that creates a familiar feel-good atmosphere.

Project details

▶**Design:** DIA - Dittel Architekten
▶**Homepage:** https://di-a.de
▶**Area:** inside area of 100 m2 and another 70 m2 outside
▶**Location:** Düsseldorf, Germany
▶**Photographs:** Martin Baitinger

About us

In our architectural office, architects, interior architects and communication designers work together to make unique creations a reality. For every project, we go through an exciting creation process that is subject to our quality standards and that, in symbiosis with inspiration, well-grounded knowledge and teamwork based on trust, results in successful project implementation.

Many years of theory and practice have made us experts in every single phase of an architectural project. In conjunction with our digital knowledge and deep understanding of our industry, we are able to provide our clients with holistic advice and design. All this makes our architectural office a place where expertise and diverse experience of our team members come together. Great teamwork is characterised by respect, transparency and understanding. It forms the basis of promising work and excellent results. This applies to working with our clients and partners as well as to cooperative work within our architectural office.

▶**Contacts** e-mail: info@di-a.de / + 49 (0)711 46 90 65 – 50

Parallels can also be drawn between the locations in Stuttgart and Düsseldorf in terms of the interior layout, which reflects the unmistakable character of Eduard's Bar. The narrow floor plan in the spatial structure is a feature of corporate design. Material and color contrasts underline the modern spatial concept. Zoning is created through the dark area in the back, which sets the perfect stage for the bar counter, and the brightly designed area for guests along the glass facade.

This area offers guests comfortable seating and the opportunity to sit in the front row. The long window façade allows sunlight to hit every spot, providing natural lighting of the interior and transforming the bar into the perfect place for select snacks and quick breakfast options during the day. The impressive panorama and the opportunity to observe what is happening along the riverside promenade from up close make Eduard's a popular meeting place.

The massive bar counter with a concrete look forms the centerpiece of the interior and is a crucial part of the unique spatial concept of the bar. The clear lines of the liquor shelves behind the bar counter pick up on the pattern of the ceiling installation. The leather-covered bar stools in combination with the brass filigree legs fit in stylishly with the overall ambience and convey a warm welcome to guests.

Another impressive feature that will surely attract visitors' attention is the selection of spirits, which underlines the distinctive bar concept and whose high quality and elegance contribute towards shaping the identity of Eduard's Bar. A height gradient of the seating areas from the low lounge area along the façade to the high bar stools at the bar counter enables visual contact between the interior and outdoor areas.

The spacious outdoor area in front of the bar, with around 120 seats, offers the perfect setting for the finely crafted lounge chairs and tables and continues the design concept from the interior. The combination of outdoor seating with a mobile bar invites visitors to sit back and relax, while the large, decorative olive trees give the whole area a Mediterranean flair.

The combination of outdoor seating with a mobile bar invites visitors to sit back and relax, while the large, decorative olive trees give the whole area a Mediterranean flair.

The reflection of the warm red colors in the windows of the bar adds an inviting ambience to the place.

Beaubourg Brasserie

Manhattan, NY, USA

classic French and new, Bar & Restaurants

Beaubourg Brasserie: Brookfield Place, 225 Liberty St. Manhattan, NY 10281, USA / +49 211 5664146660

Beaubourg
· BRASSERIE ·

Located inside Brookfield Place, Le District is a French shopping district and restaurant destination bringing an authentic Parisian experience to downtown Manhattan. The massive culinary hall is comprised of four distinct districts—offering freshly made crepes and pastries, a dedicated French fry counter, grab-n-go food stations, a wine bar, a curated retail area with groceries and fresh produce, multiple eateries, and interactive wine and cheese classes from in-house experts. There's also L'Appart, a hidden Michelin-starred fine dining experience led by chef Nicolas Abello. Basically, this place has everything imaginable to satisfy all your French food and drink cravings.

Beaubourg, the brasserie that is the flagship dining space of the market, has a slightly mod look, and a generally softer vibe than the rest of the place. Its adjoining bar, Le Bar, is a little bit darker and sexier, and equipped with a stylish neon sign (soon to be the focus of many Instagrams, no doubt). The market is opening section by section. So far, the cafe and pastry areas, and the candy shop from French company La Cure Gourmande, as well as Beaubourg and Le Bar are open.

The restaurant captures this same casual and creative energy with its seasonal menus; as each menu changes its product mix, the cover design changes drastically and is themed after a French artist or movement. The rest of their visual identity, from marketing collateral to their web presence, feels artsy and free.

Project details

▶ **Design:** Crown Creative
▶ **Homepage:** www.crowncreativenyc.com
▶ **Area:** inside area of 230 m2 and another 10 m2 outside
▶ **Location:** Manhattan, NY, USA
▶ **Image:** @crowncreativenyc

About us

We are a tight-knit team of creatives who believe that close relationships and collaboration are the keys to effective brand building. We are a studio of designers, writers, artists, and strategists built to create brands from the bottom up. We believe in the sum of the parts; the idea that creating a brand is a journey. It's not only just design, but a set of ideas, beliefs, values and actions we take to get there. Everything we do starts with a clear strategy. We listen to you, pinpoint what makes your brand special, and develop a clear plan.

We have our finger on the pulse of hospitality and design, and fuse our intuition with the best artistic talent to produce beautiful, thoughtfully-crafted identities that look, sound, and behave like your brand. We strive for functionality and believe that the emotional impact of a space is just as important as the aesthetic. Designing mobile-first, responsive websites to showcase your brand and services. We are Squarespace & Shopify experts with over 100 built websites under our belt.

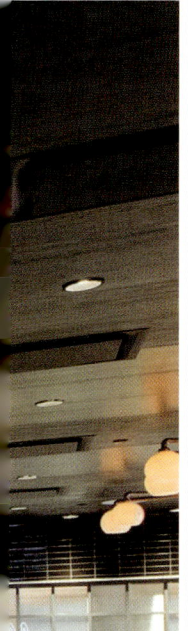

Le District's fine dining location, Beaubourg Brasserie is both classic French and new New York. Chef de cuisine Alexandre Petard provides a seasonal menu of dishes for breakfast, lunch and dinner. Ingredients are pulled fresh from the market and combined in traditional and modern recipes, making any reservation a dining experience to remember.

The restaurant & Bar captures this same casual and creative energy with its seasonal menus; as each menu changes its product mix, the cover design changes drastically and is themed after a French artist or movement. The rest of their visual identity, from marketing collateral to their web presence, feels artsy and free

Beaubourg
MENU EVENTS RESERVATIONS LE BAR GALLERY CONTACT
Beaubourg
BOTH CLASSIC FRENCH, AND NEW NEW YORK
RESY BOOK NOW
MacBook

beaubourg.com
Beaubourg
beaubourg.com
Beaubourg
Events
WHETHER YOU'RE PLANNING A CASUAL COCKTAIL HOUR OR A FANCY DINNER, CAPTURE THE MOMENT WITH US AT LE BAR OR BEAUBOURG.
Explore crafted cocktails, an extensive wine list, and decadent French-inspired dishes and hors d'oeuvres from the Beaubourg kitchen. Let our savoir-faire cater to your event needs; leave the details to us.
Name *
beaubourg.com
Beaubourg
Brunch
Served Sat & Sun from 8am – 3 pm
- HORS D'OEUVRES -
Soupe a L'Oignon 12
Gruyere Gratin, Sour Dough
Steak Tartare* 19
Hand Cut Filet Mignon, Truffles, Quail Egg, Baguette
Country Frisee Salad 16 / 19
Bacon Lardons, Bleu Cheese / Add Truffled Poached Egg

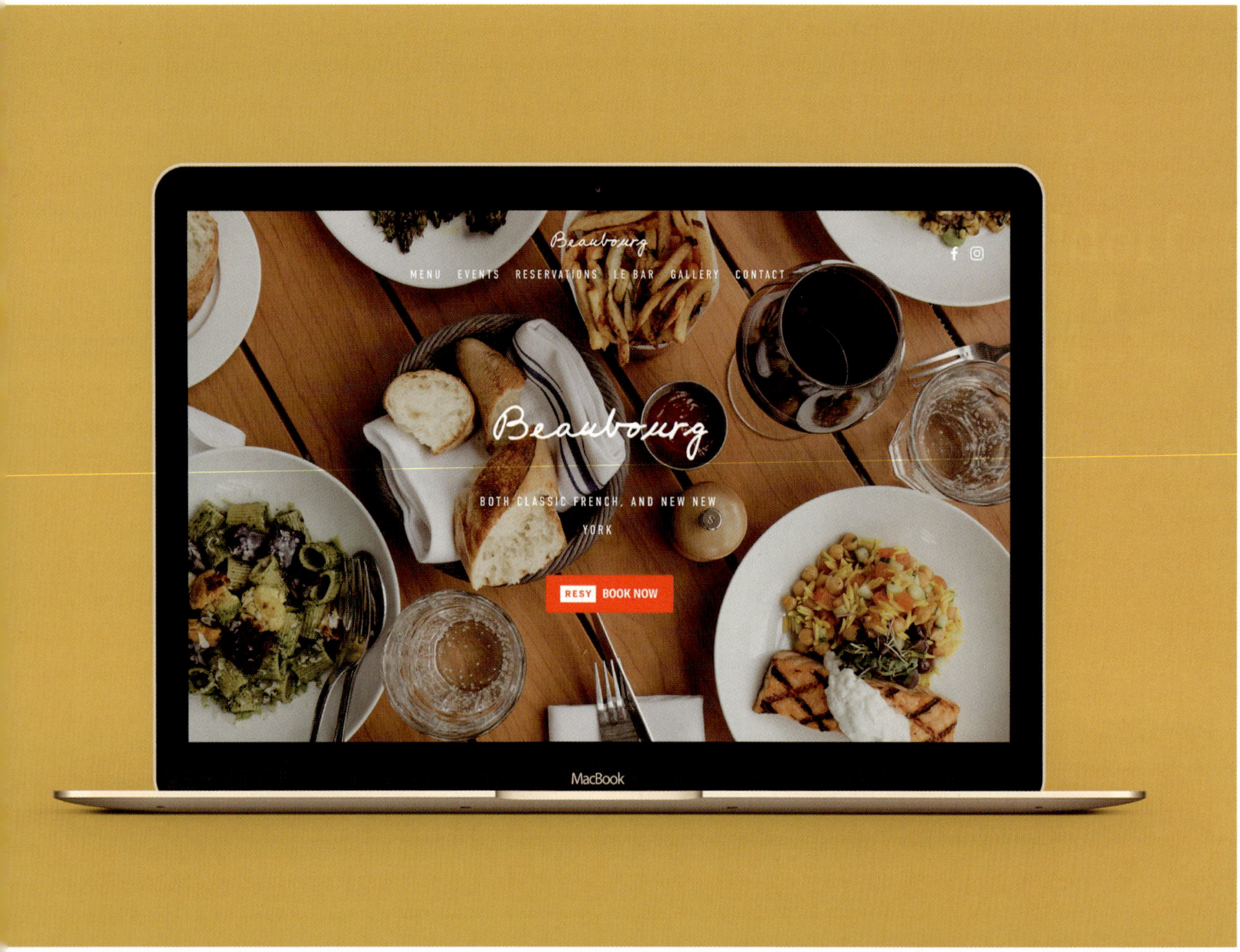

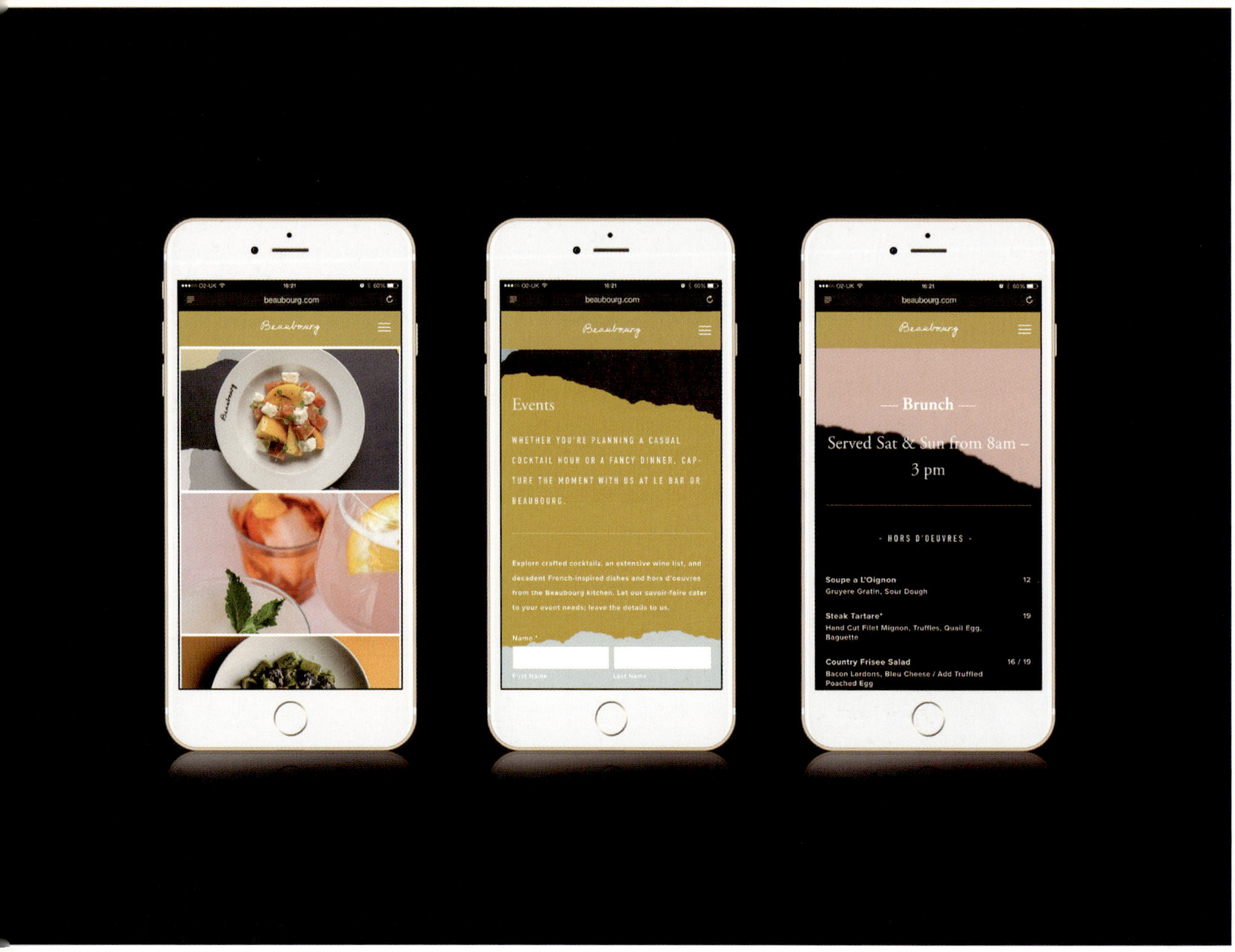

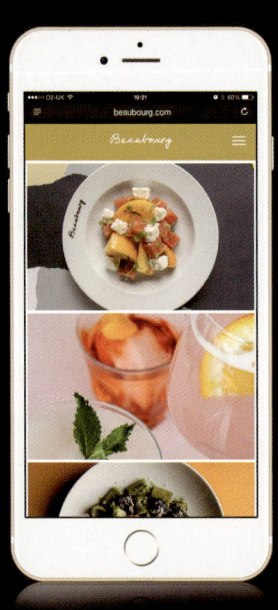

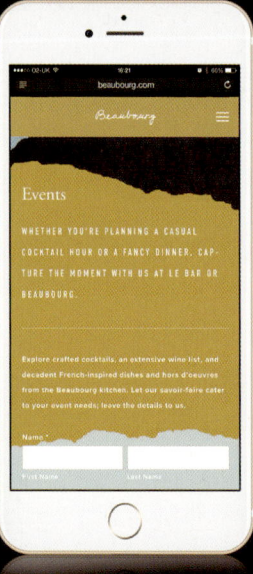

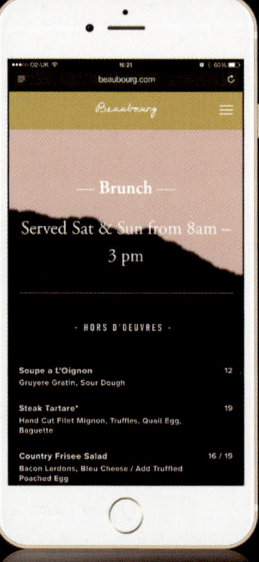

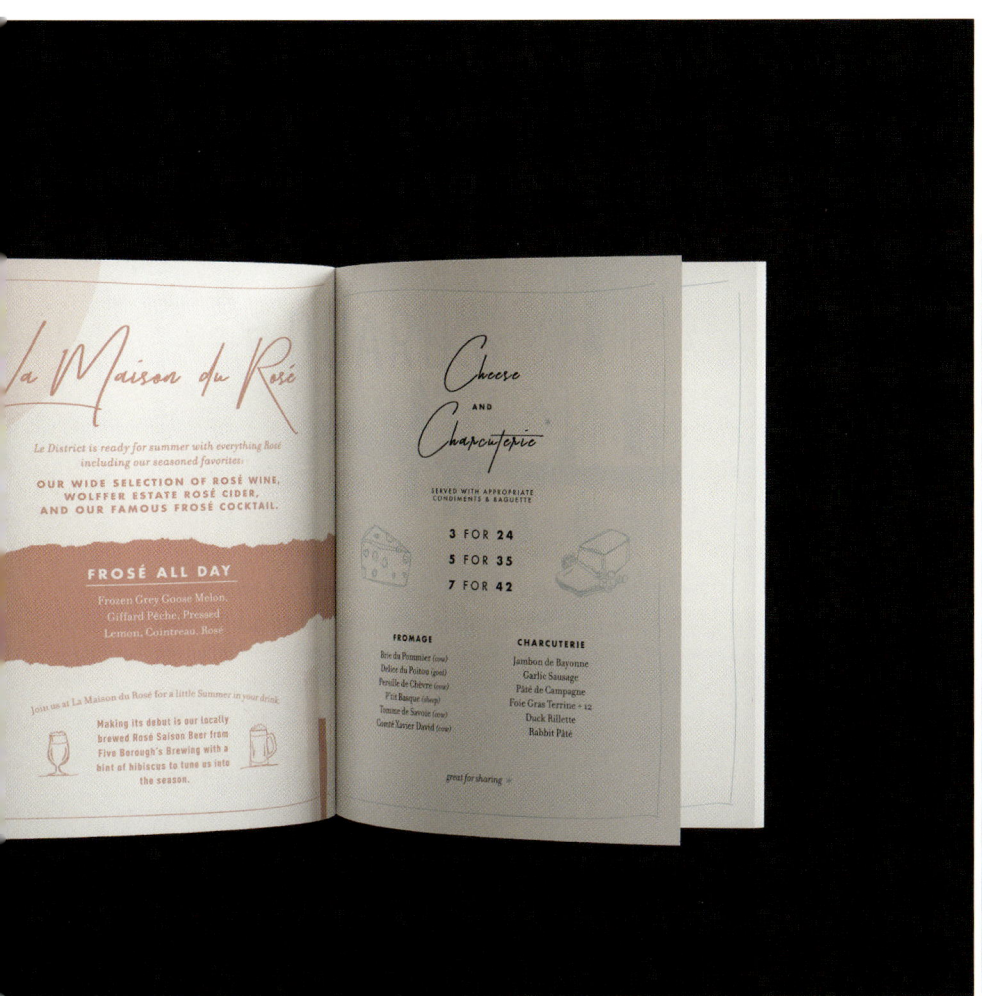

The flagship dining space of the market, has a slightly mod look, and a generally softer vibe than the rest of the place.

The massive culinary hall is comprised of four distinct districts—offering freshly made crepes and pastries, a dedicated French fry counter, grab-n-go food stations, a wine bar, a curated retail area with groceries and fresh produce, multiple eateries, and interactive wine and cheese classes from in-house experts.

Kiev downtown, Ukraine
Punkraft bar

Ater brings the rebellious spirit of craft beer to downtown kiev.

Ater architects has designed a new bar located in the podol area of downtown kiev in ukraine. formerly home to a cocktail bar, the design of 'punkraft' intends to pay homage to the rebellious nature of the craft beer movement. the architects were tasked with the challenge of changing the schematic layout of the premises, equipping it with a cold room for storing beer kegs and completely redesigning the interior, including the entrance and guest zones.

The resulting design by the architect utilizes materials that represent the industrial aesthetics of beer equipment. stainless steel, cold rolled metal, concrete and neon lights are used to create a rebellious atmosphere in the bar area. in contrast, softer and more tactile materials are applied in the main guests' areas to provide a more comfortable ambience. this is achieved by using oak wood, textile upholstery and warm muted light. in addition, the original, historical brick arches have been retained in the interior.

The plan of the space is divided into three zones: the bar, communal areas and quiet areas. the main feature of the bar is a 3.5-meter-long, 24 tap, custom draft beer tower connected with the cooling chamber. it allows a constant rotation of assorted craft beers on tap at all times. the ceiling above the bar is made up of metal arched segments as a reference to the brick vaults of neighboring halls. overall, the design of 'punkraft' effectively invites craft beer lovers with it's industrial, punk interior.

The metal arched segments above the bar refer to the historic brick vaults. Arched screens have been used to emphasize the archlines, the two aisles of the communal zone are specially designed to host large groups, the quiet room furnished with cozy sofas invites for more peaceful and relaxed socializing. A niche hand-painted by alexander grebenyuk, a renowned ukrainian street artist. Aak bars with the name of beer, its style, alcoholicity, the percentage of bitterness and cost. the walls have been decorated with a pattern of street tags.

Project details

- **Design:** Ater architects
- **Homepage:** https://www.aterarchitects.com
- **Area:** 215 sq.m
- **Location:** Kiev downtown, Ukraine
- **Photography:** Andrey Avdeenko

About us

Founded by alexander ivasiv and yuliya tkachenko, studio based in Kyiv, Ukraine. We Provide architectureal and interior design solusions with complete supervision from concept to realization. Worldwide.

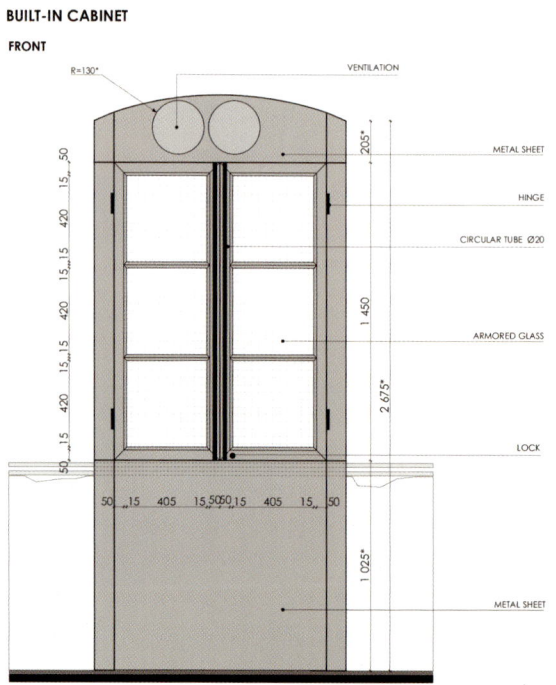
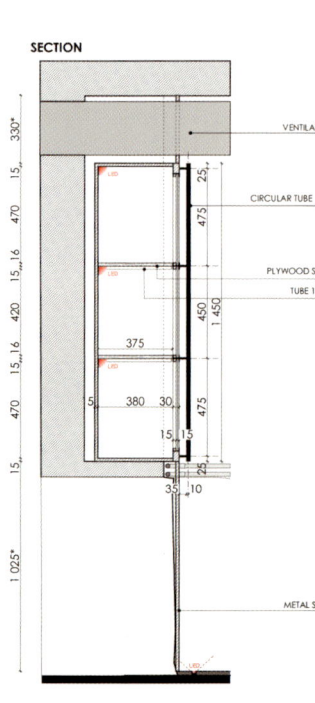

BUILT-IN CABINET

FRONT — SECTION

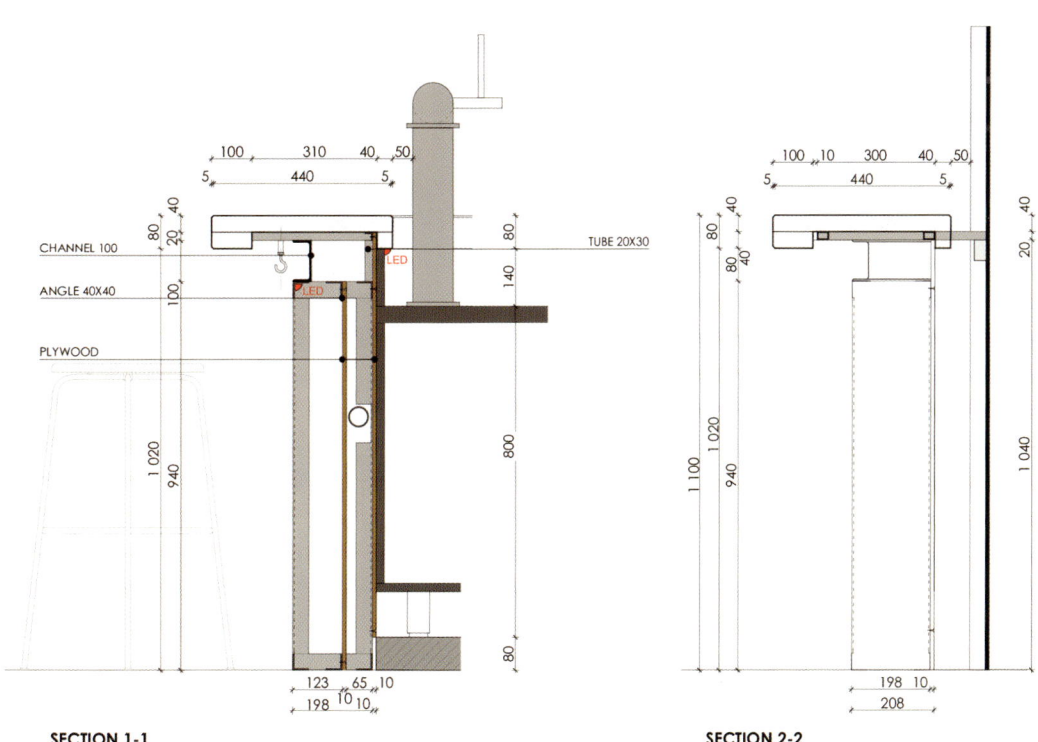

SECTION 1-1

SECTION 2-2

BEER TOWER

FRONT

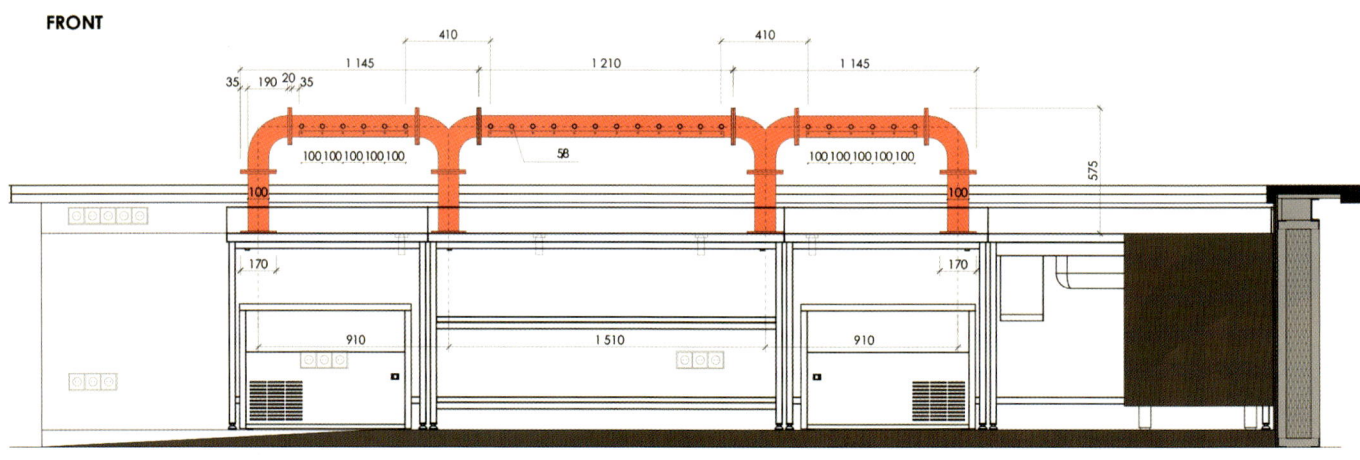

TOP

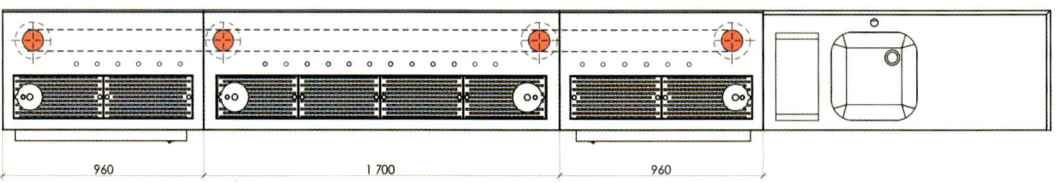

BAR ZONE

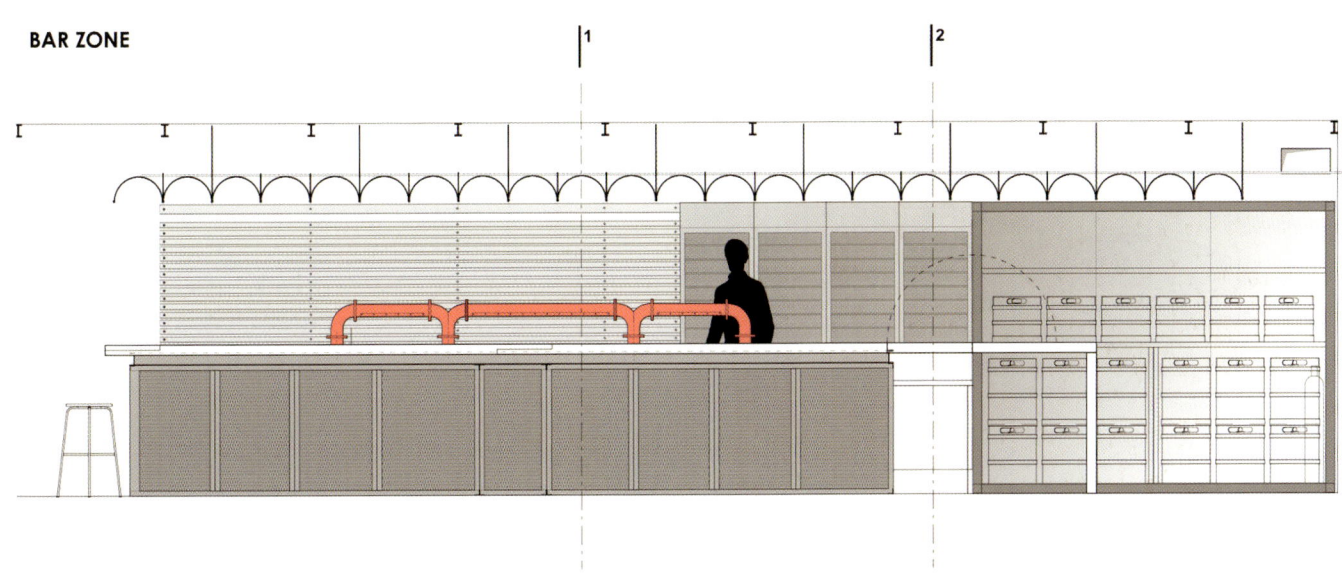

NO SMOKING
KEEP DRINKING

18+ NO ALCOHOL FOR KIDS.
PLEASE TRY AGAIN AFTER
YOU'VE REACHED 18

BAR SEATING ZONE

CONSOLE TABLE

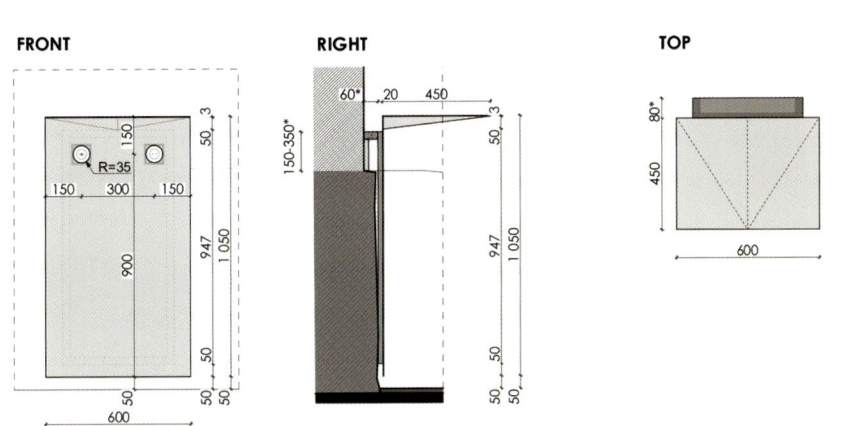

FRONT **RIGHT** **TOP**

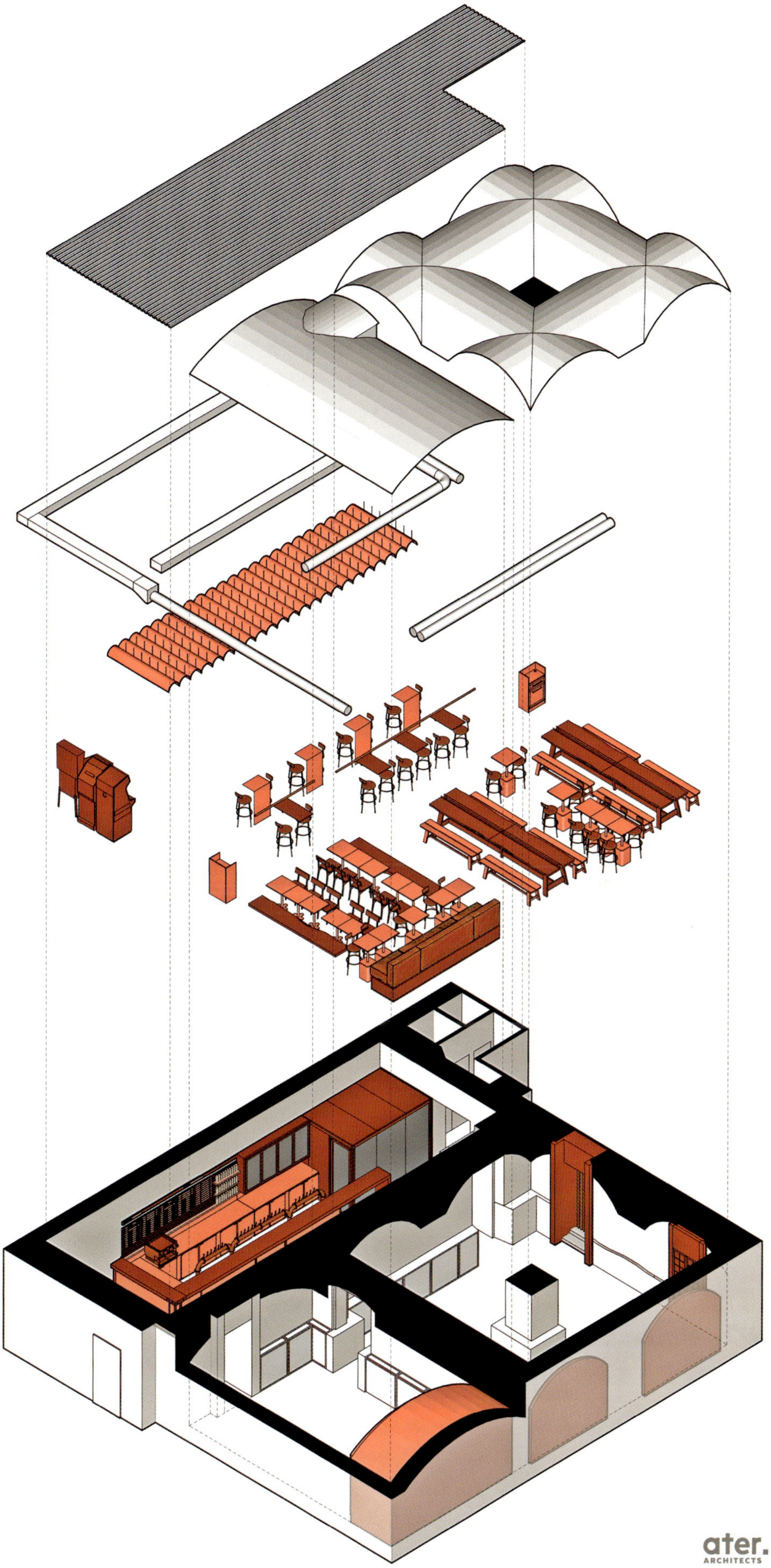

095

A niche hand-painted by alexander grebenyuk, a renowned ukrainian street artist. Aak bars with the name of beer, its style, alcoholicity, the percentage of bitterness and cost. the walls have been decorated with a pattern of street tags.

COMMUNAL ZONE
QUITE ZONE
KITCHEN
ENTRANCE
FRIGE
BAR
WC
WC

Floor Plan

WINS

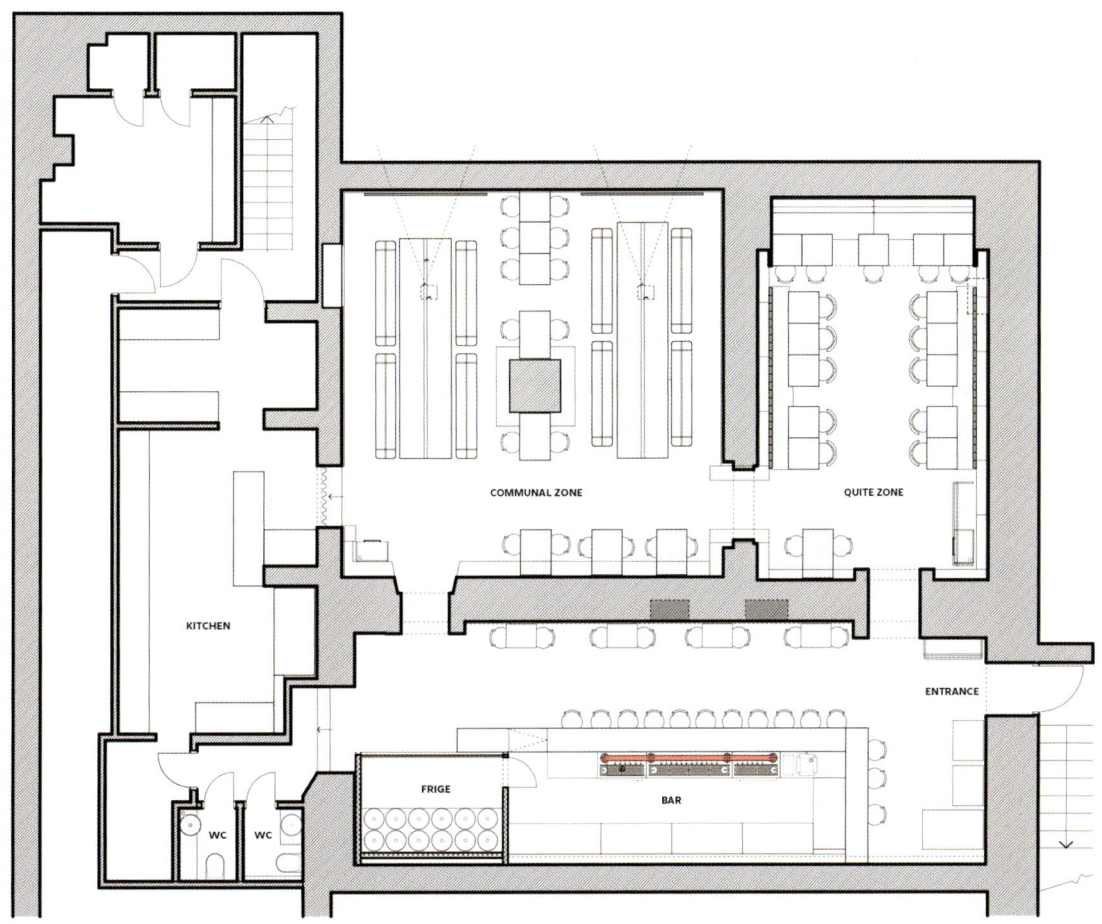

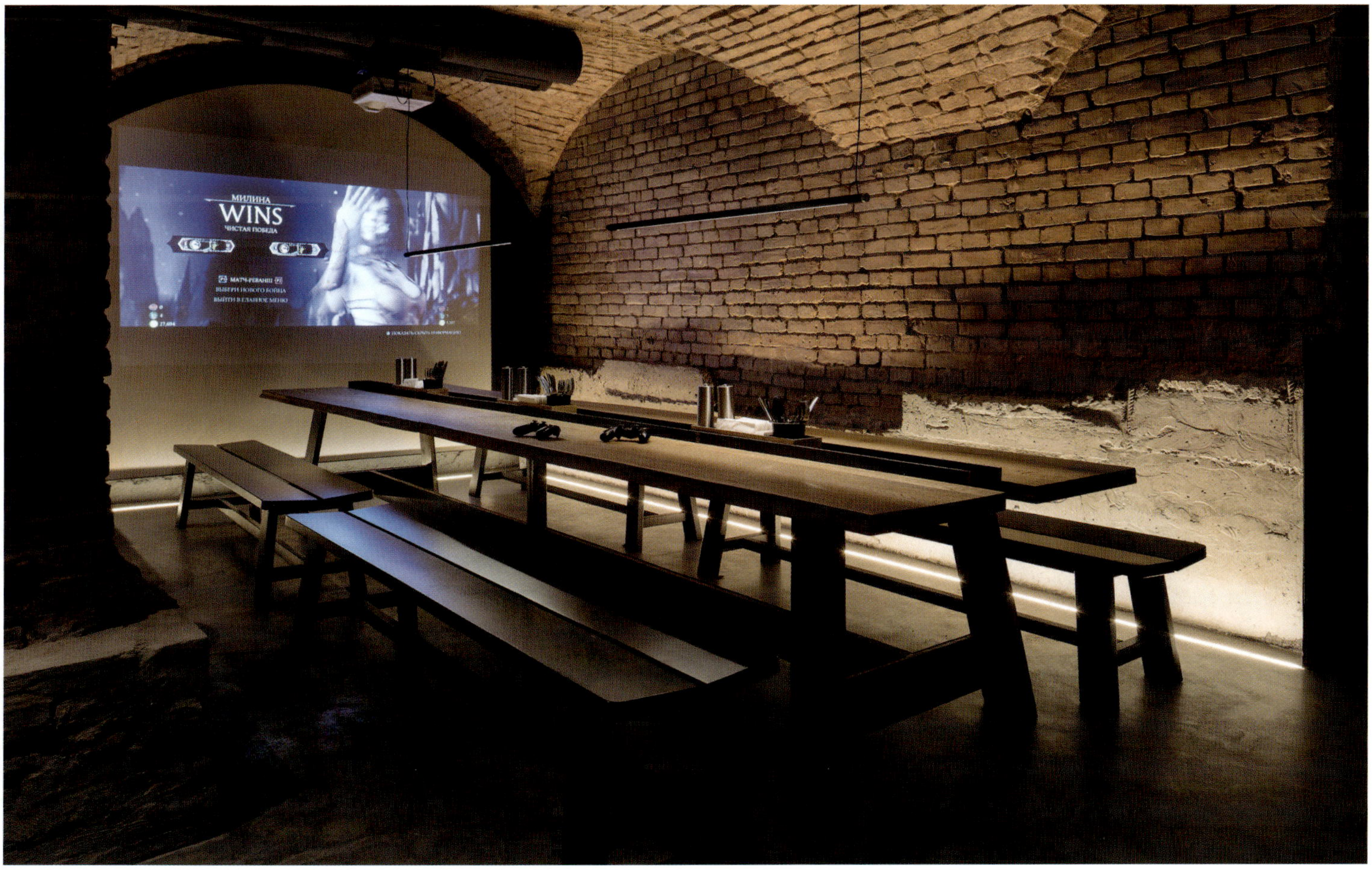

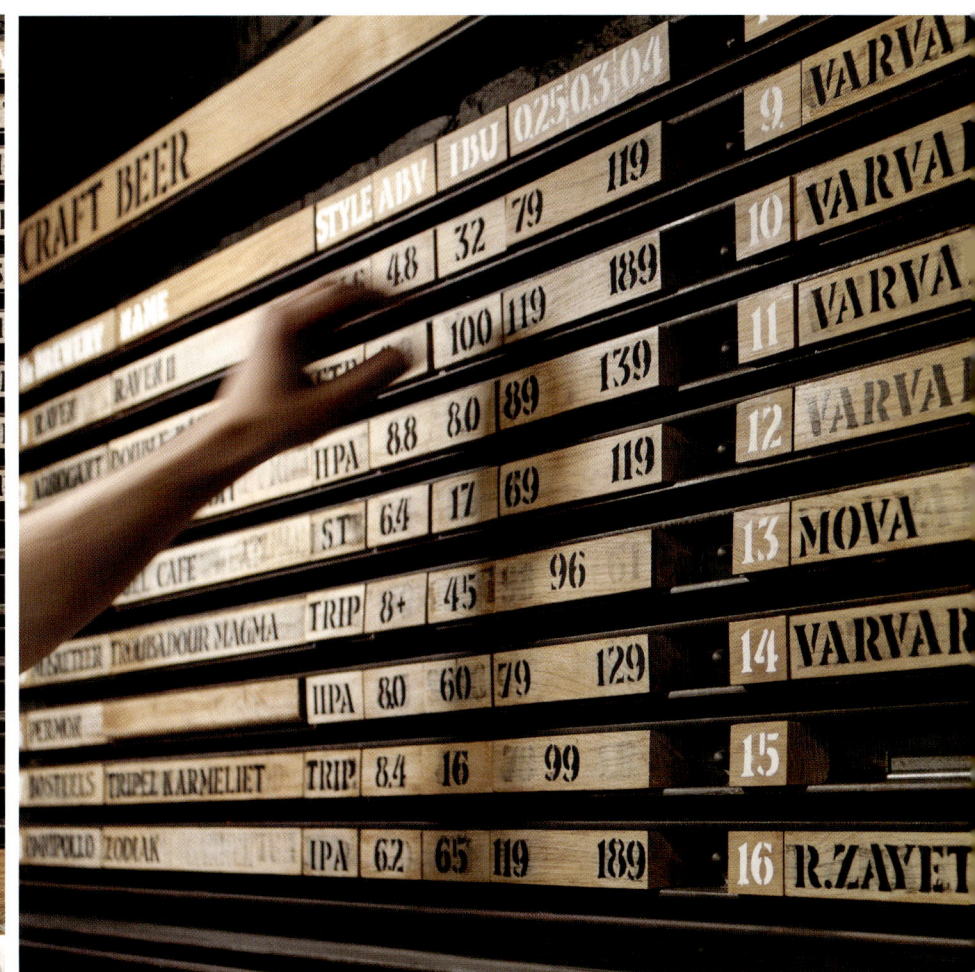

BAR MENU

FRONT

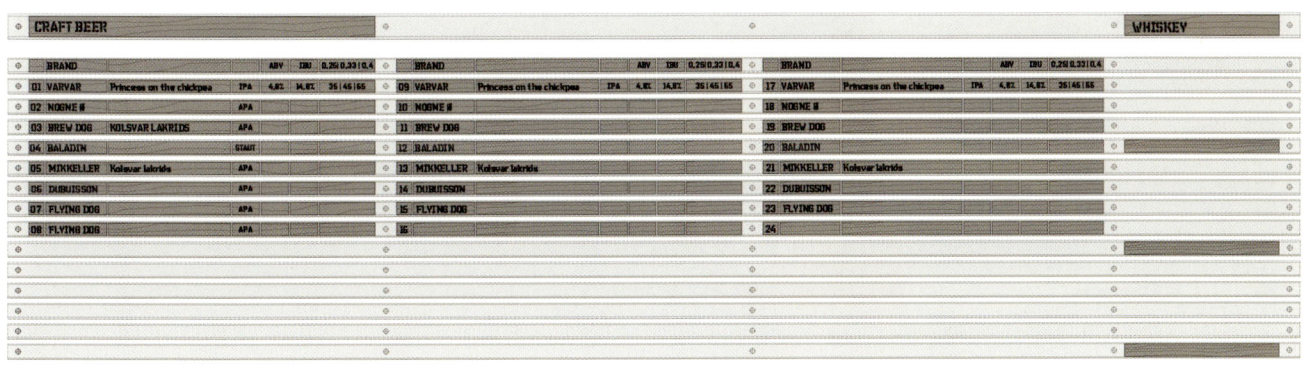

RIGHT

SECTIONS

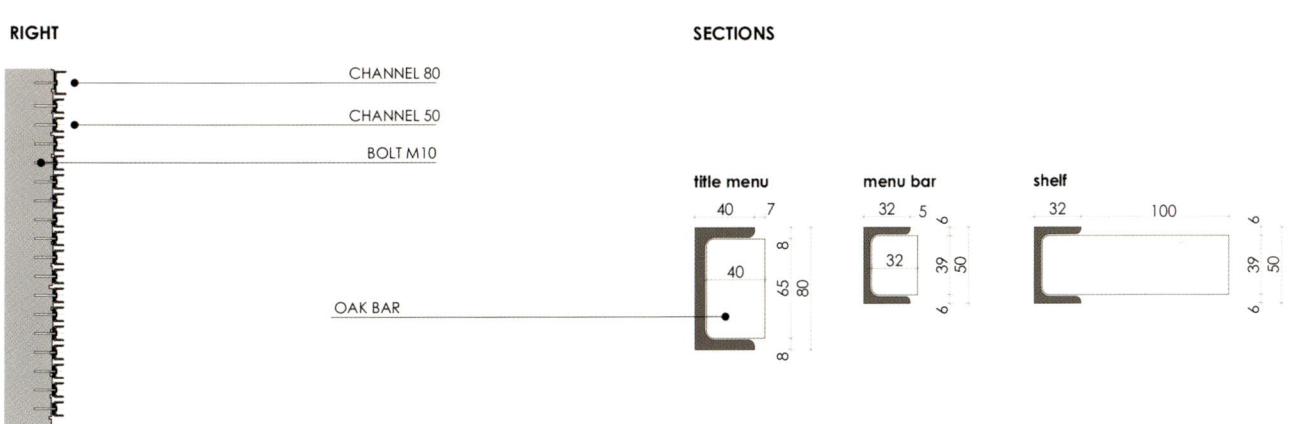

Shenzhen, China
Tarentum Bar

MARS created a cavernous dune restaurant in Shenzhen's Futian central business district amid rows of tall buildings. The high-end brand Tarentum Restaurant of Shenzhen Ludejin Catering is located on the central axis of Futian District Center of One Avenue, facing the hollow square stairs, close to the main entrance of the B1 commercial area, and closely connected with the urban public space.

The design starts with the connection between the entrance space and the sunken plaza. Despite the complex outline of the plan and the limited size of the space, MARS still decided to leave the entrance area to the city and connect with the public space of the sunken square. Through four pivot glass doors, the design introduces the urban public space into the interior and brings people more sense of belonging.

Due to the unreasonable planning of the original structure and equipment of the building, the clear height of the interior is not uniform: the highest point is 4.2m, but the lowest point is only 2.8m. The challenge was to make the most of the existing interior height and to integrate the Spaces clearly and uniformly. MARS designed a series of dune-shaped arched ceilings that serve as a link between different heights. This solves the problem of indoor height and creates a unique art space like a cave.

Project details

▶**Design:** MARS Studio
▶**Homepage:** https://www.mars-studio.com
▶**Area:** 230 m²
▶**Location:** Shenzhen, China
▶**Photography:** Wen Studio

About us

MARS Studio was founded in New York, and opened its Beijing office. It is an international architecture studio full of vitality and energy oriented towards architecture and interior design.

MARS attaches great importance to every link from project planning, conceptual design to project construction, committed to the perfect pursuit of design quality, attach great importance to the design of engineering details and the real completion of the design, attach great importance to the real needs of owners and the unified relationship of architectural expression of social significance. Therefore, we will be responsible for on-site supervision of each project.

The L-shaped bar at the main entrance connects the indoor and outdoor spaces and turns the experience from public space to private dining. The corridor leading to the dining area is also carefully decorated with surprises: at the beginning of the hall is a niche for sculpture; at the end of the corridor is a culture wall written in Greek. In the dining area, with the undulations of the dunes, the spatial attributes change from open to semi-enclosed to private, dividing different levels of space.

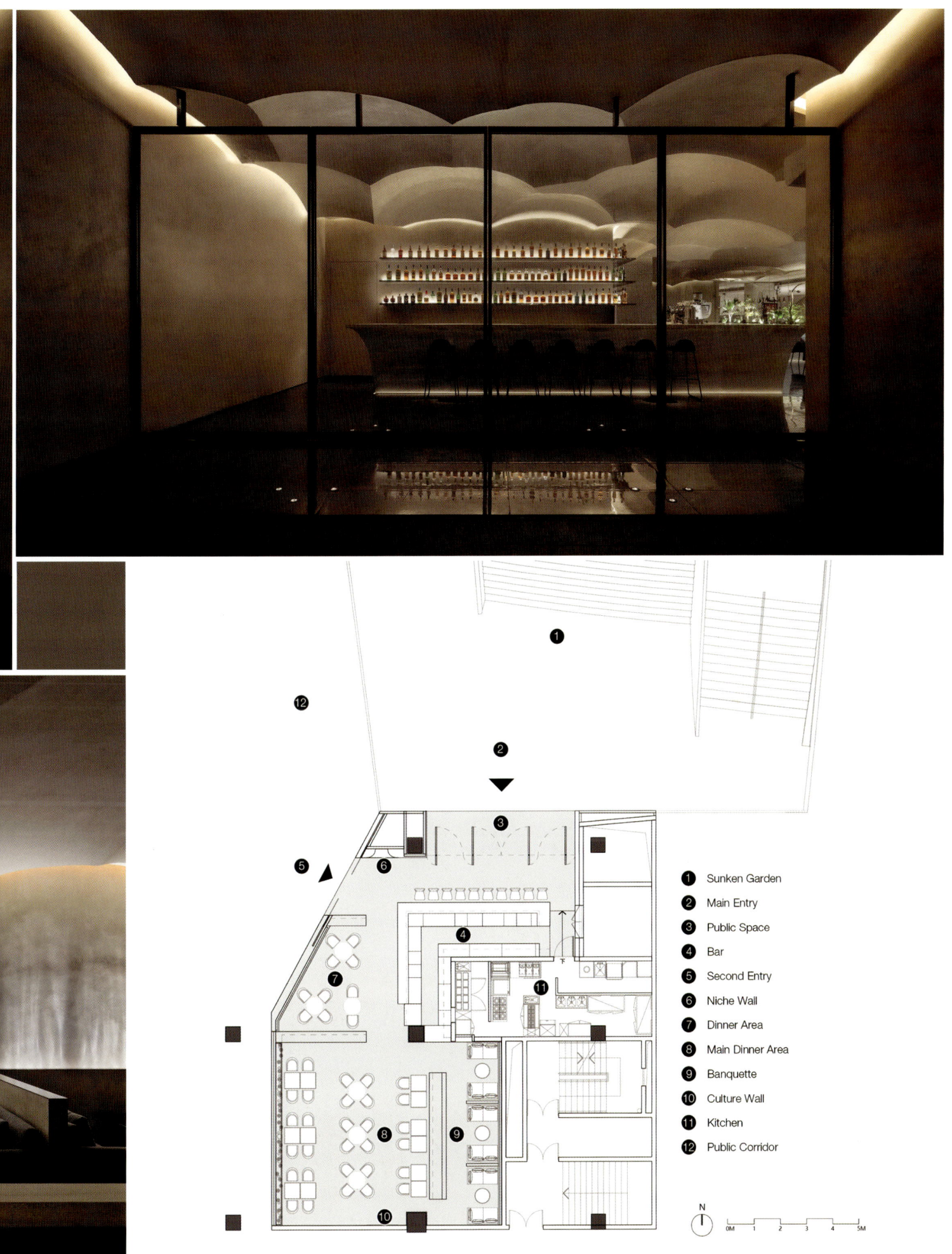

1. Sunken Garden
2. Main Entry
3. Public Space
4. Bar
5. Second Entry
6. Niche Wall
7. Dinner Area
8. Main Dinner Area
9. Banquette
10. Culture Wall
11. Kitchen
12. Public Corridor

The selection of materials in the interior focuses on the subtle contrast under the overall unity: the straight wall is deliberately "dirty" texture paint, and the curved top is uniform clean micro cement. The straight linear light along the junction of the wall and the ground echoes the curved light arranged along the surface of the ceiling, which jointly sets off the lighting atmosphere of the wall and the ultimate dining experience.

MARS designed a series of dune-shaped arched ceilings that serve as a link between different heights. This solves the problem of indoor height and creates a unique art space like a cave.

The renovation of the facade an "inside-out" design. The "dune" extends from the interior to the exterior, becoming the main visual point of the sunken square. Tarentum bar & restaurant breaks through the inherent boundary between commercial indoor space and urban public space, inadvertently attracting people in this busy city of Shenzhen to stop and enter the sand dune space to relax, rest, and find themselves.

Tarentum bar & restaurant breaks through the inherent boundary between commercial indoor space and urban public space, inadvertently attracting people in this busy city of Shenzhen to stop and enter the sand dune space to relax, rest and find themselves.

The straight linear light along the junction of the wall and the ground echoes the curved light arranged along the surface of the ceiling, which jointly sets off the lighting atmosphere of the wall and the ultimate dining experience. The renovation of the facade an "inside-out" design. The "dune" extends from the interior to the exterior, becoming the main visual point of the sunken square.

Kyiv, Ukraine
Bar Marokana

Handcrafted tiles and glass bricks with warm lighting create a portal of entry into the mysterious city.

▶Bar Marokana : boulevard Lesi Ukrainky, 24, Kyiv, Ukraine / +380 67 481 9202

As is known, Europeans have always loved to explore the culture of exotic countries. Traditions of architecture, crafts, and colour schemes have had a huge impact on creative people for a long time. These gave rise to the emerging of colonial architecture and interiors in the homeland, as well as the increase in trend reversal - the emergence of the cultural object directly in distant lands. For example, brightly coloured oases with light motifs of France or Belgium can be guessed, and local cultural features are respected.

The proximity of fashion boutique SPAZIO was essential in creating the interior of Morocco. White minimalist, sterile interior fuses with a warm amber room designed by loft buro. The Fashion design would have to get a logical line in the interior; hence Yves Saint Laurent and his beautiful villa in Morocco happened to be a source of inspiration. Hand-crafted tiles and glass bricks with warm illumination create the portal of the entrance to the mysterious city; as a reminder of the magnificent cities along the ocean shore, on the "edge of the earth".

The natural, nude colour and terracotta's material bear the memory of the touch of the master's hands and dominates the interior, covering floors, walls, and even ceilings. Pergolas (sun protection structures) on the ceiling give an impression of being in the villa. Chandeliers with plants embody the tradition of terracing greenery, and illumination - the eternal sun.

Project details

▶**Design:** Loft buro
▶**Homepage:** https://loftburo.com/
▶**Area:** 180 m²
▶**Location:** Kyiv, Ukraine
▶**Photography:** Andriy Avdeyenko

About us

Loft buro, est. 2001, is a creative team, which consists of professional architects, designers and painters.

During this period of time, many projects in interior design and architecture were made by the team. Our main task is the creation of a harmonious, comfortable and cosy space that displays the inner world of a person and creates the perfect mood for all to enter it.

Handcrafted tiles and glass bricks with warm lighting create a portal of entry into the mysterious city.

The walls with large mirrors symbolize the colour of the Atlantic Ocean, which washes the sandy shores. Glass bricks were created especially for this project and integrated into real brickwork. Amber light pouring through them reminds the sun`s rays.

The combination of terracotta with many shades of Azur blue and deep blue, some brass as a symbol of gold, are the main range of the interior. As for the bar`s stand, the enlarged relief element in glazed ceramics was created using the traditional ornaments.

The glossy structural surface magnetically draws attention to itself and is the main focus of the space. The Eastern classic lamp lines were used in the shelves` design on both walls of the bar area. The brass elements resemble garlands of lanterns strung on the axle.

Ceramic countertops, chairs, sofas are made in the style of handmade artisans. Natural and relict wood, ropes, carpets with deep shades, mild earth colour of all materials emphasize the aesthetics of a country house.

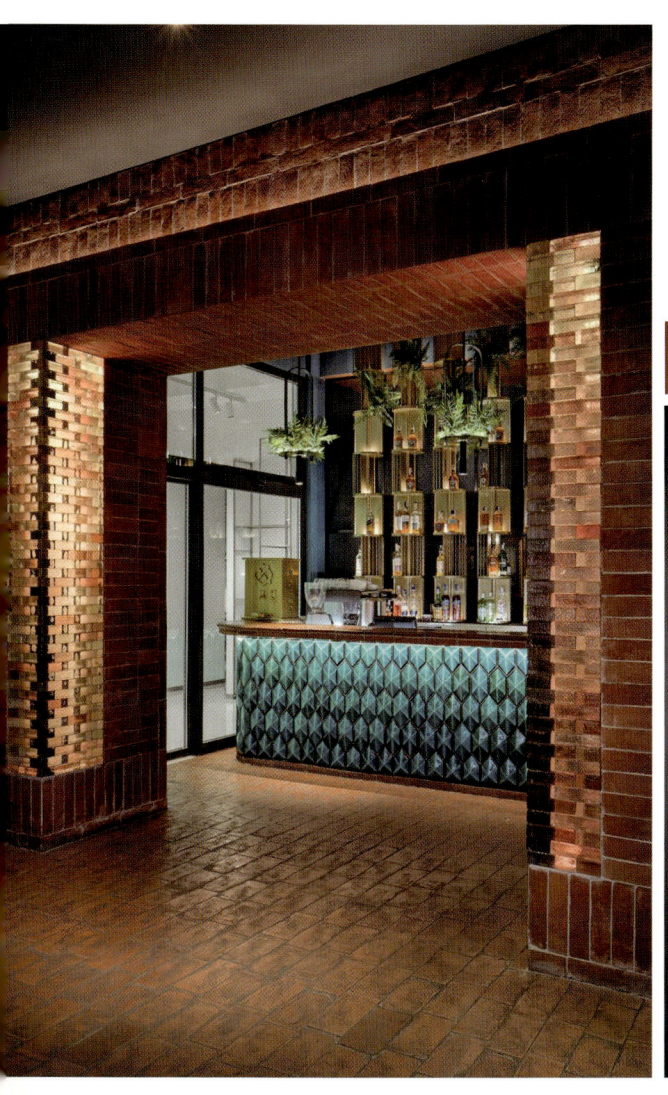

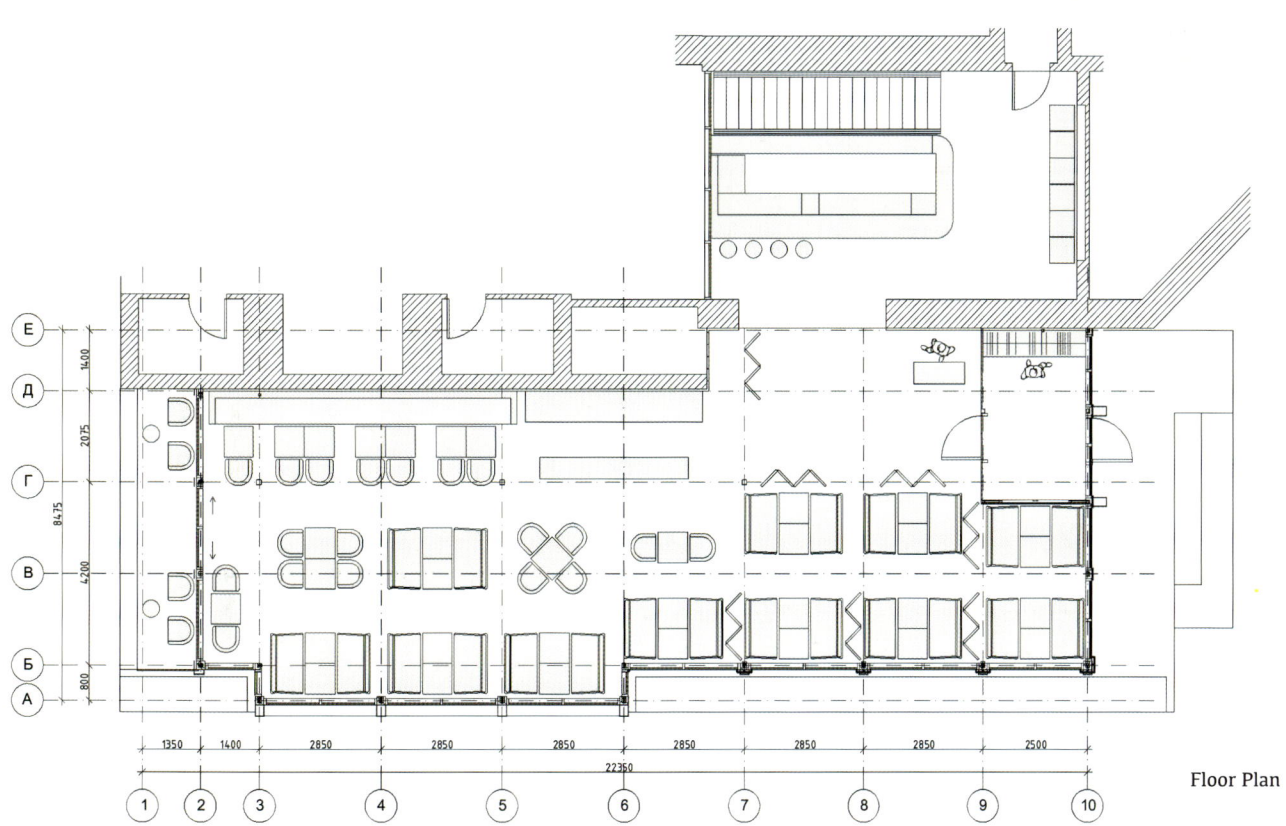

Floor Plan

Ceramic countertops, chairs, sofas are made in the style of handmade artisans. Natural and relict wood, ropes, carpets with deep shades, mild earth colour of all materials emphasize the aesthetics of a country house.

Furthermore, Inga Makarova`s paintings depicting exotic animals, aboriginal women promote the impression of an artist`s workshop or a collector`s villa.

The mysterious city; as a reminder of the magnificent cities along the ocean shore, on the "edge of the earth".

If you want to taste Moroccan meals, take your chance and visit this restaurant. You can order good pilaf, fish and couscous. Take your chance to taste delicious wine. You will enjoy great coffee or good tea at MAPOKAHA.

The combination of terracotta with many shades of Azur blue and deep blue, some brass as a symbol of gold, are the main range of the interior. As for the bar`s stand, the enlarged relief element in glazed ceramics was created using the traditional ornaments.

127

Rabbit Habit bar: Shota Rustaveli St. 24, Kyiv Ukraine / +380 98 024 0024

Kyiv, Ukraine
Rabbit Habit bar

On the 7 P.M., the restaurant from the healthy food changes to an easy-bar

The venue is located in the central Kyiv, in the basement of the building erected in the early XX century. You would not miss it because a huge rabbit's head above the entrance is noticeable from a long distance. The restaurant consists of two halls with a bar counter in each of them. Visitors can observe a cooking process in the open kitchen. There are 46 seats over there, one big table for 8 people and a terrace.

The main idea of Rabbit Habit's interior is to combine nature and the rhythm of a city life. There is a lot of timber in the interior different in its shades and textures. Table-tops have an imitation of bark beetles, the main bar counter built from wooden blocks has a beautiful pattern of squares in its end. You can see massive timber elements on the entrance. They were brought from the Carpathian region where they used to serve as beams in an old house a long time ago. Small stone dust on the floor, in WC, and on tabletops on terrace makes the interior as touchable as only possible.

The symbol of the venue is a wooden rabbit's sculpture cut of ash. They are everywhere – the biggest one is placed next to the register, middle rabbits stay on the counter and shelves. A waiter brings a check to a client pinned to a metal plate by a tiny wooden rabbit with a magnet inside. On the wall and sound protective panels on the ceiling, you can see a pathway of a rabbit's footprints.

Project details

- **Design:** YOD Group
- **Homepage:** https://yoddesign.com.ua
- **Area:** 134 m²
- **Location:** Kyiv, Ukraine
- **Photography:** Andriy Bezuglov

About us

Our story began in 2004, when studio's founder Volodymyr Nepyivoda and art director Dmytro Bonesko united their efforts and brought together talented architects and designers. Today we work predominantly in the sphere of commercial design and create projects in hospitality sector: hotels, restaurants, cafes and bars.

Taking on any new project, we want to become a reliable partner who is familiar with restaurant business and shares Client's vision. First of all, we search the idea which will be borne in people's mind and bring long-standing customers to you. To create a perfect image of the object, we design individually furniture, lighting and decoration, implement experimental solutions, create naming, graphic design and dishes serving.

Today we have the best restaurants in our portfolio and successful restaurateurs and hoteliers among our Clients. Our projects are published in famous magazines and win prestigious competitions. But the most important thing is that they bring our Clients love and loyalty of the guests.

The low ceiling was one of the main challenges of the project. Designers decided not to make it lower, but arrange the air-conditioning system inside the pipes with quite a small diameter under the ceiling. Cone lamps above tables were made from the same kind of pipes.

The second hall doesn't have windows, an atmosphere there is maximum chamber-like. Rounded walls and a big installation of dried herbs on the ceiling create a relaxing mood. There is a salad-bar which transforms into a bar-mixology at 7 P.M.

On the terrace, you can sit on the Rabbit Chairs by Stefano Giovannoni and Qeeboo. They are highlighted from inside and make this area noticeable in the evening Kyiv's street. High steel tables with green herbs planting in the center keep the general idea of being close to nature.– On the 7 P.M., our rabbit turns to his ears, and the restaurant from the healthy food changes to an easy-bar. The atmosphere is changing as well. The evening lighting scheme is cozier and chamber-like. The interior we created works well in both scenarios. – Volodymyr Nepiyvoda, co-founder of YOD Design Lab tells.

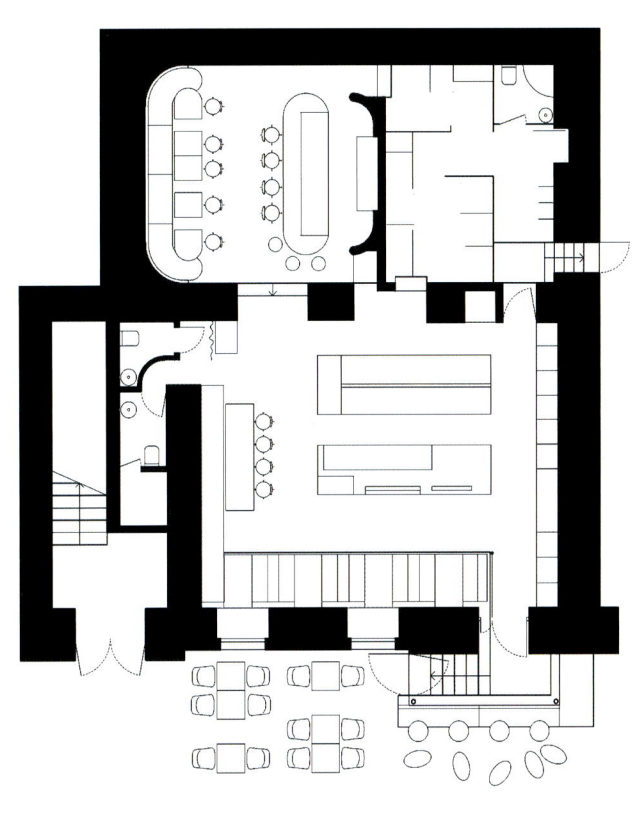

Floor Plan

Rabbit Habit is located in the basement of a building on the city's Shota Rustaveli Street. With limited ceiling heights to work with, and the need to add ventilation to the restaurant, the designers, YOD Group, decided to use this exposed pipeworks to create cylindrical lamps above tables.

The second of the bar's two main spaces features a striking art installation, made up of dried herbs, flowers and grasses that appear to be growing out of the ceiling overhead to haunting effect. Timber wall panels further the natural feel while creating an acoustic chamber for more intimate gatherings. Counters and furniture have been built using salvaged blocks and planks of timber – fashioned from historic Carpathian house beams. Rabbit motifs also dot the space, peering out of planters, perched next to tables and lined up outside the bar front to add a playful touch to the Kyiv bar.

The restaurant's menu includes bowls, salads, avocado toast, soups, meat and fish dishes, as well as sugar-free desserts from Svitlana Savchenko and superfoods.

The second hall doesn't have windows, an atmosphere there is maximum chamber-like. Rounded walls and a big installation of dried herbs on the ceiling create a relaxing mood. There is a salad-bar which transforms into a bar-mixology at 7 P.M.

Orvay wine bar
Barcelona, Spain
Barcelona's Orvay bar takes design cues from winemaking

Orvay wine bar: Pg. del Born, 4, 08003 Barcelona, Spain / +34 938 32 45 04

The surrounds of Orvay lend themselves to dramatic flair. The wine bar is set in front of the Santa Maria del Mar, one of Barcelona's most emblematic 14th Century Gothic churches. The interiors follow suit, with the art of wine-making serving as the inspiration behind Isern Serra and Sylvain Carlet's design. Colours of the land, vine, and grape are at the fore, the contrasting shades used to denote three various seating sections. "We liked the chromatic relationship that was generated between the three spaces so there is no element that separates them visually," says Serra and Carlet. "We wanted colour to be the great protagonist of the space." In the entrance bar, rounded white marble high tops give a sculptural contrast to the building's medieval bones. Serra and Carlet knocked through a false ceiling to reveal historic arches and stone walls. The natural, earthy tone represents the Denomination of Origin, the official geography used to identify the land on which grapes for different wines are grown. The reflection of a round, merlot-hued central mirror is meant to give the effect of looking through a glass of red wine.

This leads through to the metaphorical vineyard; a more formal dining section with walls, ceiling, and floor seamlessly painted in the same deep shade of vine-green. A black steel wine rack houses some of Spain's best bottles, and two-person marble tables are unified by a natural oak bench that runs the length of the wall. A large mirror fills a pre-existing medieval arch, expanding the space. The back dining room, reserved for large groups, is painted pink in honour of a winemaker's essential ingredient, the humble grape. Lighting plays a key role in enhancing, but not overwhelming, the bold colour palette. Serra and Carlet are particularly inspired by the "simple geometries" of straight lines and spheres in their work. Linestra LED lights hang throughout the space, while an Anastassiades for Flos lamp takes precedence over the communal dining table.

Project details

- **Design:** Isern Serra, Sylvain Carlet
- **Homepage:** https://www.isernserra.com
- **Area:** 134 m²
- **Location:** Barcelona, Spain
- **Photography:** Jose Hevia & Salva López

About us

Isern Serra opened his own studio in the 22@ neighborhood of Barcelona in 2008, from where he carried out interior design projects, exhibition installations and industrial design. He combines his professional career with teaching, giving classes and workshops at schools like Elisava or IED. His work has been exhibited in cities such as Barcelona, Milan, Paris, Dusseldorf, Mexico DF or Beijing and published in specialized industry magazines. He has been invited as a speaker on various occasions, highlighting the invitation from ICEX to present his work at the "Beijing design week 2014". Currently, he has won the competition to carry out the new concept of a science museum for Barcelona's Cosmocaixa and the stands for the furniture companies Enea and Expormim, at the Milan Fair.

He is a very versatile designer capable of tackling projects ranging from the exhibition field, offices, shops or restaurants and his clients are both from the public sphere such as Sala Canal Isabel II, BCD, Palau Robert or private such as "laCaixa", They cover the projects from simplicity and honesty, giving birth to their own identity that allows them to give each project a unique character.

The bar – named Orvay – allows visitors to sample an extensive variety of wines, which served as a starting point for local designers Isern Serra and Sylvain Carlet to develop the bar's interiors.

It developed a colour scheme inspired by elements of wine production, picking contrasting shades that could outline different seating zones.

Decor throughout the bar has been kept to a minimum, excluding a handful of large, red-tinted circular mirrors that are supposed to evoke "looking through a glass of red wine".

Which Isern Serra and Sylvain Carlet have designed in reference to the process of making wine.

The false ceiling's in the bar's entryway – which features casual high counters and black stool chairs by Danish brand Menu – has been knocked through to reveal historic stone walls. These have been left exposed with the intention that their beige hue will remind visitors of the fertile earth of wineries. In the more formal dining area, surfaces have been painted deep green in reference to the verdant terrain of vineyards. This room has also been dressed with a long oak bench seat and tables topped with white marble.

Santo Cocktail bar
Naples, Italy

Santo! Everyone came to mind, and what better name for a concept bar in Naples.

What distinguishes a cocktail bar from a simple bar? Definitely the atmosphere in which you can enjoy your favorite drinks. Santo is a restaurant located in a nineteenth-century building in the heart of the Chiaia district in Naples, in via Bisignano, a context rich in nightlife.

The request of the clients was immediately to create a place with a captivating, elegant and fresh interior design, but which also had the distinctive features of the Neapolitan architectural culture and tradition. The challenge was therefore to make these two requests coexist in a single concept between the design of the equipment and the use of refined materials. From a design point of view, also in function of the narrow and long rectangular planimetry, it was decided to contrast on the two long sides of the rectangle, on one side, floral wallpaper, green ceramic tiles in pastel tones, arched structures in electro-coloured aluminum, mirrors and led, on the other the tuff of the walls and the marble, cladding of the bar counter and of the bench positioned at the back of the room.

Two souls that coexist harmoniously facing each other within the same environment. Two sides of the same coin ideally united in the circularity of time by nature, transfigured in the ceiling of the room: a light metal lattice suspended on a black background hosting the lights and suspended plants, precisely nature. Even the custom-made equipment of the room, the bookcase and the bottle racks, are specular and complementary, as they are generated one from the other through the subtraction and translation of a shape (bottle rack) from a larger solid (bookcase). The same materials compose them but the internal finishes are different, the heart, in marble for the bottle shelves and antiqued mirrors, which always allow you not to lose track of what was once one and is now triune, for the equipped wall.

Project details

▶ **Design:** IN-NOVA STUDIO
▶ **Homepage:** http://www.in-novastudio.com
▶ **Area:** 102 m²
▶ **Location:** Naples, Italy
▶ **Photography:** Carlo Oriente

About us

IN-NOVA STUDIO was born in Santa Maria La Nova, in the heart of Naples, from the ideas and meeting of three young architects: Marcello FERRARA, Martina RUSSO and Riccardo TEO. The three, strengthened by their training and professional experiences acquired in various international studios, shared the project of returning to their homeland to deal first-hand with the issues posed by contemporary society to the architect.

IN-NOVA STUDIO searches for innovative solutions for each single project it deals with; experiments with materials and technologies taking into account the resources of the territory and the needs of the customer; carries out projects from interior design, to restoration, to architectural and urban planning, with a critical and multidisciplinary approach. IN-NOVA STUDIO designs a dynamic, functional, sustainable architecture, attentive to the historical context, the environment and the urban and extra-urban landscape, aware of the responsibility of being an architect.

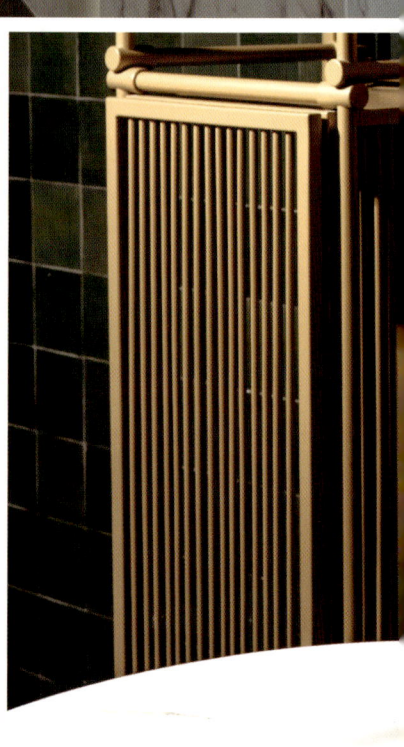

▶Mama Cocha: Piazza Garibaldi, 26, San Giuseppe Vesuviano, Italy / https://www.mamacocha.it

Naples, Italy
Mama Cocha

Mama Cocha, a journey through flavors from Naples to Peru.

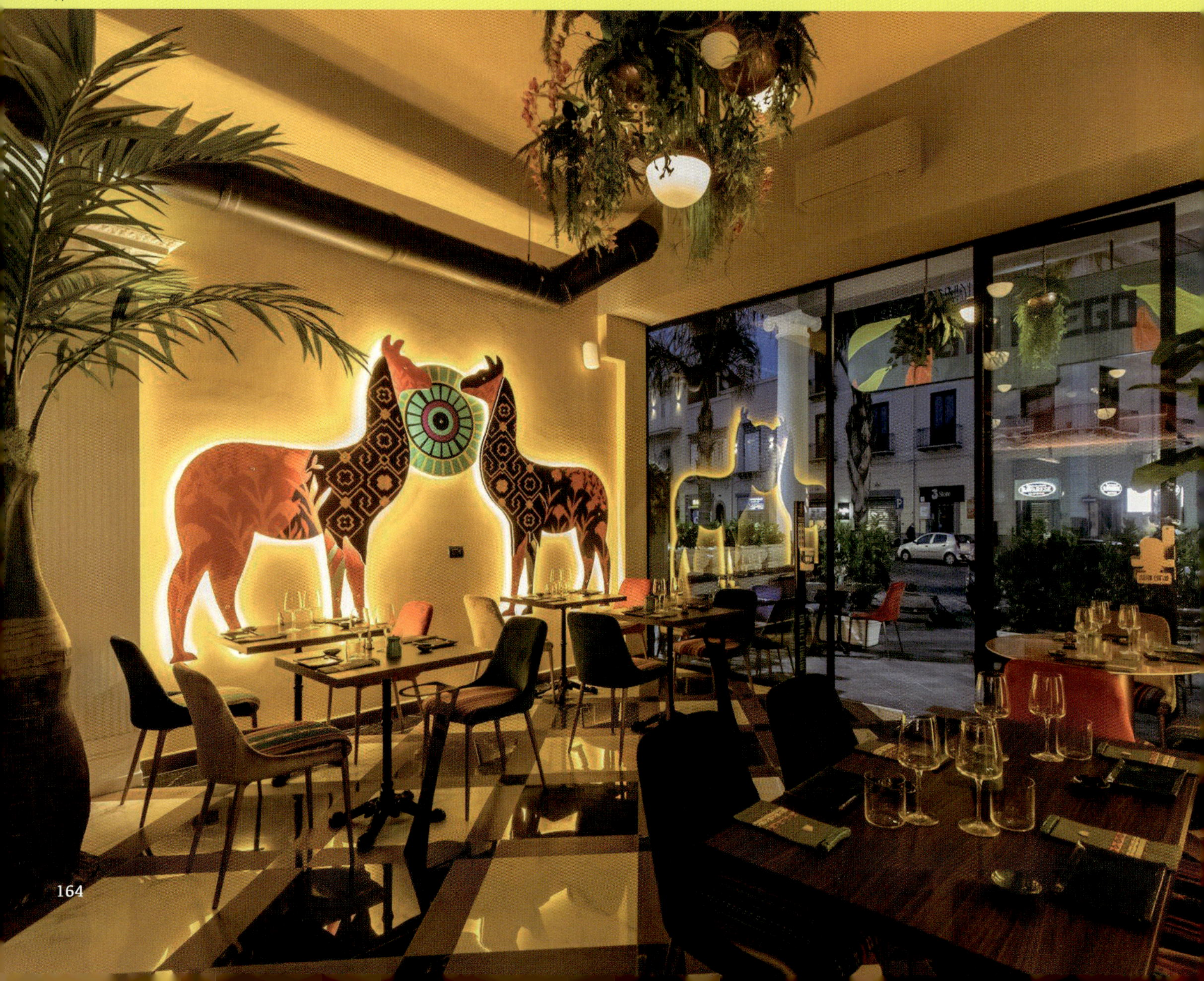

The interiors of Mama Cocha, designed by the architect Mario Sorrentino, perfectly reflect her fusion and lively soul: bright colors, starting from the sunny yellow walls and ceiling, Peruvian decorations reinterpreted in a contemporary key, cascades of plants that embrace chandeliers and boiseries and customized furnishings.

One of the strong points of the restaurant is its location. The outdoor patio overlooking the wonderful Sanctuary of San Giuseppe offers a pretty good view, while the indoor furniture, young and in full South American style, focuses on refined contamination and projects us into a playful and sparkling world. Stylistically different materials and decorations mix with brilliant color palettes in an aesthetic riot of great effect.

Just like the Sofia seats by Riflessi, customized for the venue and dressed in the typical fabric originating from Peru, decorated with a multicolor striped pattern, combined with a velvet-effect fabric in strong colours, such as orange, ocher and blue. The base of the seats, on the other hand, focuses on the elegant rose gold finish, which adds a chic touch to the eclectic style of the venue.

The atmosphere is inviting and the culinary offer well studied. There are dishes already "tested" in the Urubamba restaurant and other new proposals, from ceviche to tacos, from steamed ravioli to classic rolls, from grilled octopus to sweet churros. For a riot of flavors that well reflect the character of the location.

Project details

▶**Design:** Mario Sorrentino architect
▶**Homepage:** www.instagram.com/mariosorrentinoarchitetto
▶**Area:** 162 m²
▶**Location:** Naples, Italy
▶**Photography:** Marco Baldassarre
▶**Furniture design:** Riflessi

About us

Riflessi customizes the seats of Mama Cocha, a restaurant in the province of Naples specializing in a nikkei food and wine proposal, a synthesis between Peruvian and Japanese culinary culture. The interiors of Mama Cocha - whose name is a tribute to the Mother of the Waters, the Peruvian goddess who protects the sea and fishing - are designed by the architect Mario Sorrentino and are characterized by a lively and contemporary fusion aesthetic, where materials meet and stylistically different decorations and bright color palettes.

For the project, Riflessi created special custom-made Sofia chairs: for the seat, a typical fabric originating from Peru was used, decorated with a multicolor striped pattern, which Riflessi subjected to a special process, combined with a velvet-effect fabric for the backrest in vivid complementary colours, such as orange, ocher and blue. A rose gold tone finish was chosen for the base of the seats.

▶Contacts www.instagram.com/mariosorrentinoarchitetto / Call us: +39 081 365 1623

One of the strong points of the restaurant is its location. The outdoor patio overlooking the wonderful Sanctuary of San Giuseppe offers a pretty good view, while the indoor furniture, young and in full South American style, focuses on refined contamination and projects us into a playful and sparkling world. Stylistically different materials and decorations mix with brilliant color palettes in an aesthetic riot of great effect.

Just like the Sofia seats by Riflessi, customized for the venue and dressed in the typical fabric originating from Peru, decorated with a multicolor striped pattern, combined with a velvet-effect fabric in strong colours, such as orange, ocher and blue. The base of the seats, on the other hand, focuses on the elegant rose gold finish, which adds a chic touch to the eclectic style of the venue.

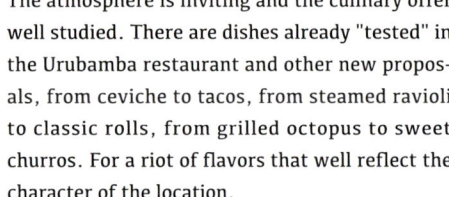

The atmosphere is inviting and the culinary offer well studied. There are dishes already "tested" in the Urubamba restaurant and other new proposals, from ceviche to tacos, from steamed ravioli to classic rolls, from grilled octopus to sweet churros. For a riot of flavors that well reflect the character of the location.

Poka Lola Cocktail Bar
Denver, CO, USA

Located in The Maven Hotel in Denver, Poka Lola reinvents the everyman's cocktail bar.

▶ Poka Lola Cocktail Bar: 1850 Wazee St, Denver CO 80202, USA / +1 720-460-2728

Located in the Dairy Block, Poka Lola reinvents the everyman's cocktail bar, with a nod to turn of the century American soda fountain culture. With its unique cocktails and eye-catching playful design, Poka Lola aims to be an inviting gathering place for all guests and occasions, whether it's a meeting destination before a night out on the town or an after-shift drink for members of the city's hospitality community.

We collaborated with SRG the Poka Lola space features an open layout with plush chairs and small tables; bold black and white-patterned floors, an elevated take on classic checkered board floors; and a stunning bar adorned by glowing stained glass panels. Along the back wall, a luminous display of antique bottles and mirrors create an inviting backdrop for a night out. In addition to playing in the bar's small arcade game area, guests will have the opportunity to enjoy cocktails in the alley come early summer.

Project details

▶ **Design:** CRÈME Jun Aizaki Architecture & Design
▶ **Homepage:** www.instagram.com/mariosorrentinoarchitetto
▶ **Area:** 1950 sq. ft. interior, 760 sq. ft. exterior
▶ **Location:** Denver, CO, USA
▶ **Photography:** Andrew Bordwin, Adam Larkey

About us

Founded by Jun Aizaki, CRÈME is a collaboration of dynamic, international designers and creative professionals. Based on the idea that all design challenges require the same problem solving approach, we approach a chair, a restaurant, a building, the same way we would approach a logo or a block. Collaboration is a key to our process. We nurture a culture of design democracy and draw inspiration from our clientele, our design team's diverse backgrounds and from our extended family of artists and fabricators. We believe in having a hands on approach to problem solving and that fresh ideas are born when moving the hand.

We take a holistic approach to design where every element speaks to each other creating a strong sense of a whole. We are storytellers and experience builders in a sense that our exploration does not begin nor does it end at the completion of a building, space or an object, but rather when a person interacts with the space and the experience enters someone's lifestyle.

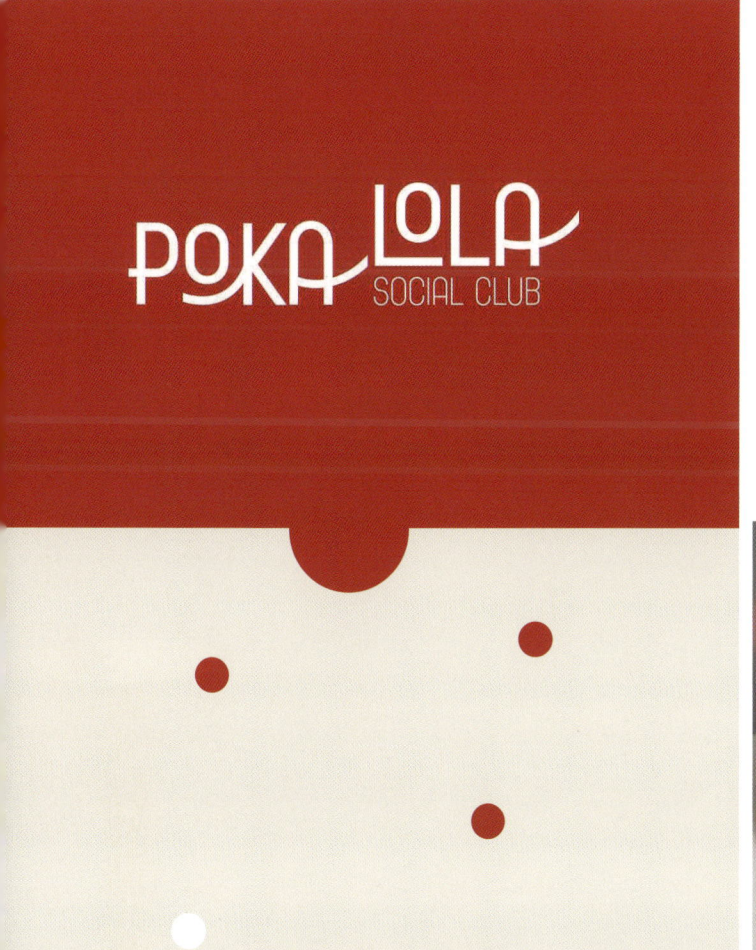

With a nod to turn of the century American soda fountain culture and a modern voice, the Poka Lola brand identity is a unique blend of old and new. Featuring vintage beer ads and extravagantly sarcastic fake reviews, Poka Lola exudes a playful and cheeky vibe. Business card sleeves feature creative cut-outs inspired by fizzy cocktails, and drink menus pull pattern inspiration from decorative stained glass above the bar.

The Poka Lola brand identity is a unique blend of old a

Business card sleeves feature creative cut-outs inspired by fizzy cocktails, and drink menus pull pattern inspiration from decorative stained glass above the bar.

With its unique cocktails and eye-catching playful design, Poka Lola aims to be an inviting gathering place for all guests and occasions, whether it's a meeting destination before a night out on the town or an after-shift drink for members of the city's hospitality community.

PAPA Dubai: Atrium Level 3, Al Habtoor City - Business Bay - Dubai, United Arab Emirates / +971 52 203 3434

Dubai, United Arab Emirates
PAPA Dubai

PAPA will take its visitors on an extraordinary culinary and mixology journey through **nine glorious bars**

The PAPA Dubai is an exciting 1,580 square meters dining and entertainment destination in the vibrant Atrium complex, the dining and entertainment epicenter of riverside Al Habtoor City. PAPA will take its visitors on an extraordinary culinary and mixology journey through nine glorious bars, each themed with a unique concept. Comfort and soul food are at the very heart of the dining experience. The fascinating bars have been developed in collaboration with Moscow and Dubai's top bartending figures.

The transformational venue transforms from a sophisticated and immersive dining experience to a chic nightlife venue with DJs and entertainment as evening moves to night. Move from a quiet dinner with friends to one of the city's best vibes. Feel the energy of this unrivaled sensory-driven experience in Dubai. PAPA Dubai will host some of the world's most courageous bartenders vying to show off their avant-garde skills throughout the year. Each of the nine bars has been carefully considered and designed by inimitable 4SPACE in a collaboration with Papa's founder Natalia Freys.

Concept

PAPA Dubai has nine very distinctive bars, each with different offerings. The Bar Village is made up of little 'boulevards' including the Rum Station, Champagne Avenue, Gin Point, And Vodka Lane, Tequila Road, Mezcal Street, Sake Alley, Wine Square, Whiskey Square, and a VIP Lounge called High gate.

Project details

▶ **Design:** 4 Space Interior Design
▶ **Homepage:** https://4space.ae/contacts
▶ **Area:** 1,580 square meters
▶ **Location:** Dubai, United Arab Emirates
▶ **Photography:** Anas Al Rifai

About us

Originally established in Damascus in 2001, founders, Firas Alsahin and Amjad Hourieh, moved their practice to Dubai to be at the centre of this vibrant market.

The emirate's booming growth in the commercial sector was an impetus for the firm to explore all the opportunities in the design industry.

Overcoming an uphill battle, 4Space Design has gone on to create noteworthy projects in the UAE. Eschewing quantity for quality, profile of the project and relationship with clients, the studio credit its people's distinct ideas strategic business development.

The Entrance
- The Entrance has an impressive, eccentric entrance with red pipe and greenery installation that hangs from the ceiling and customized carpet below.

Various design elements

To create a cohesive village-feel, archways in different finishings are used throughout the expansive space to have a unified design between each of the bars. The monochromatic color approach allows a seamless transition between each distinctive bar.

"VIP LOUNGE"
"CHAMPAGNE AVENUE"
"DJ BOOTH"
"GIN POINT"
"VODKA LANE"
"TEQUILA ROAD"
"MEZCAL STREET"
"SAKE ALLEY"
AV CONTROL ROOM
SHISHA ROOM
"HIGH GATE"
DANCE FLOOR
"VIP LOUNGE"
"WINE COURT"
SCREEN
FEMALE TOILET
"RUM STATION"
HVAC ACCESS
TOILET PASSAGE
BOH ENTRANCE
"WHISKEY SQUARE"
RECEPTION AREA
MALE TOILET
OFFICE AREA
DRY WALL PARITION HEIGHT 2.5 METERS
BOH AREA
ELEC. ROOM
HABTOR PALACE ACCESS
ALCOHOL STORAGE/BAR LAB
LAB AREA
SHAFT
TELEPHONE ROOM
ELEC. ROOM
KITCHEN AREA
HVAC ACCESS
PLAN
SCALE 1:200

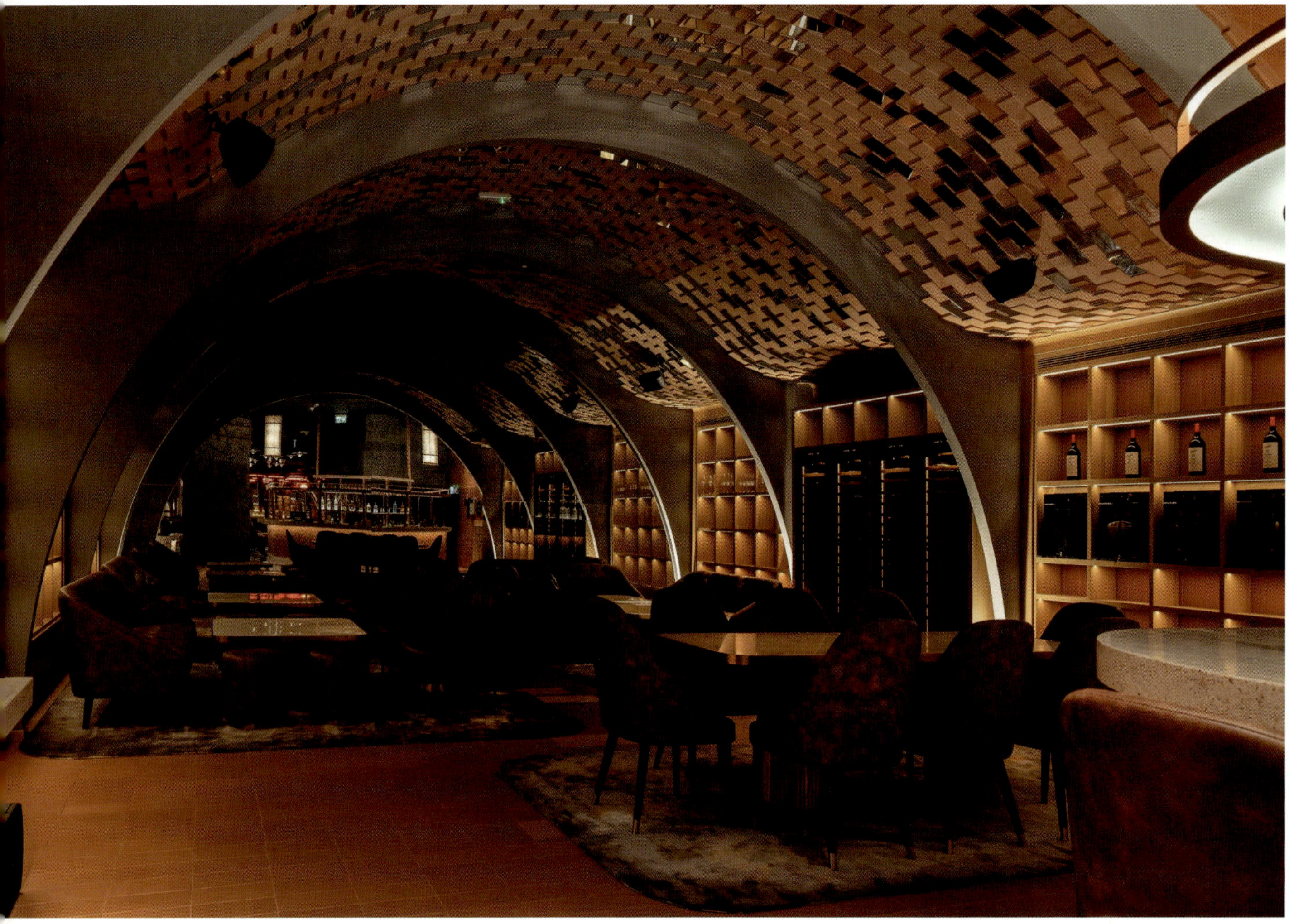

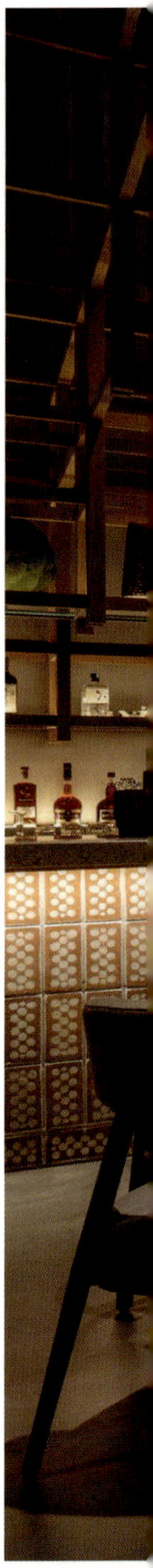

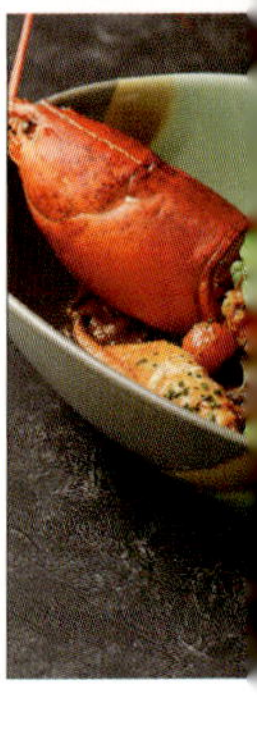

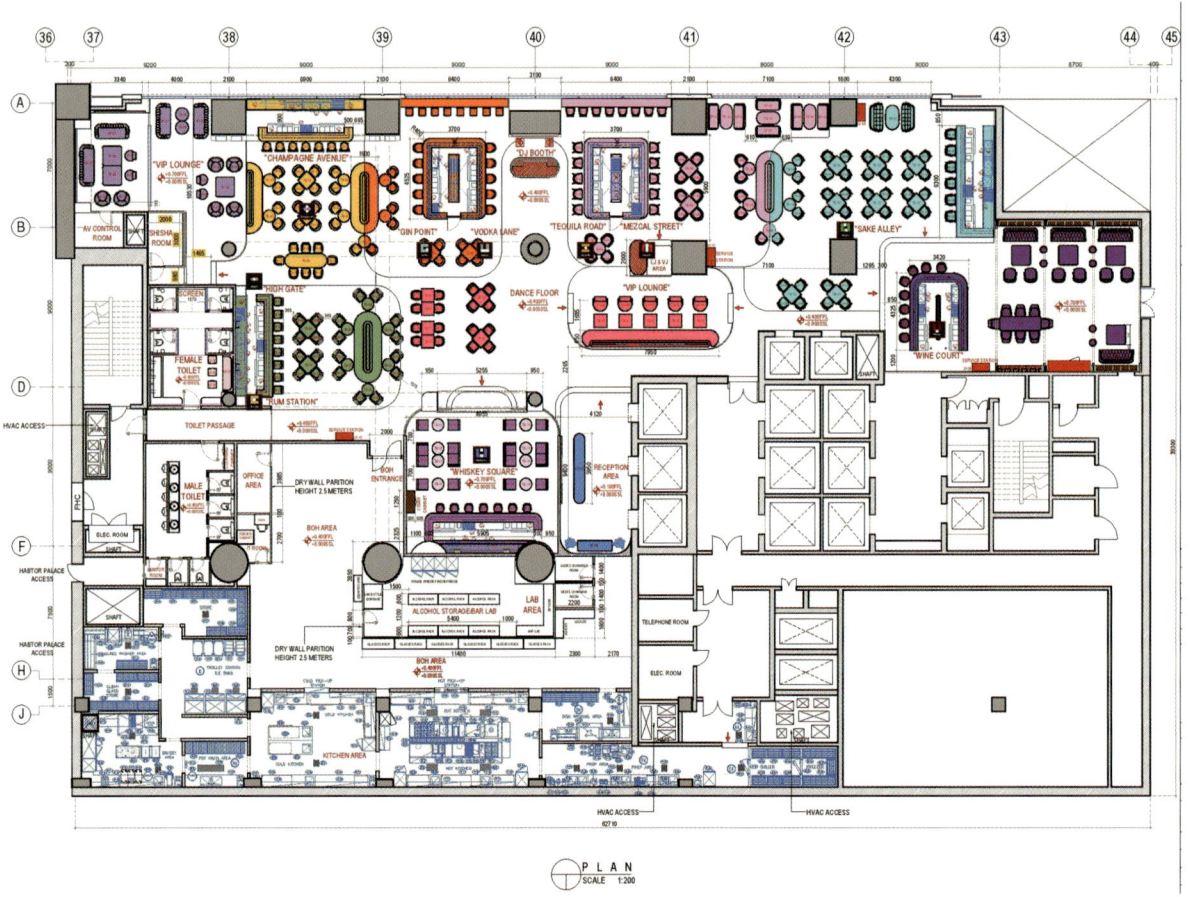

Unique key features - Wine Court
- Designed as a wine cellar with four prominent arches and an oversized mirror to reflect the arches into eight
- Terracotta on the walls and floors
- A mixture of bricks and mirrors on the arched ceiling
- Copper details inset into the bar and furniture

About the execution
4SPACE allowed two months for the design and another five months for the fit-out and transformation. There were various challenges because of the sheer variety of materials required for each bar, the complex mechanical, electrical, and plumbing (MEP) requirements, and the acoustic challenges that take the bars from soulful eateries to exciting night venues.

About the ambiance
Each bar has its own distinctive vibe and personality and transforms daily from a chic restaurant where you can dine in style as the evening draws to a close. As nighttime falls over the city the restaurant metamorphoses into an elegant party venue. The VIP Lounge is the final destination of the night.

Challenges overcome

There were several challenges that 4SPACE had to overcome. Eight different terrazzo surfaces were designed and installed throughout Papas. Cleverly, designers sourced authentic materials and graffiti artists to allow for fully cohesive individuality. 4SPACE presented the clients a 3D design which they executed to perfection. The piece de resistance is the nine thoroughly different concepts within one narrative. The arches unify the overall design.

Unique key features - VIP Lounges
- There are two VIP Lounges for an elevated experience
- One is red and serves as an extension to the entrance, the brand colour of PAPA and boasts three oversized lighting pendants
- he other is centered with an enormous arch and copper mesh on the wall, black marble on the floor, and gorgeous lounge seating

Unique key features -DJ Booth
- A space that can easily convert into a dance floor because of its entertainment lighting and circular kinetic mirrors, and led lighting

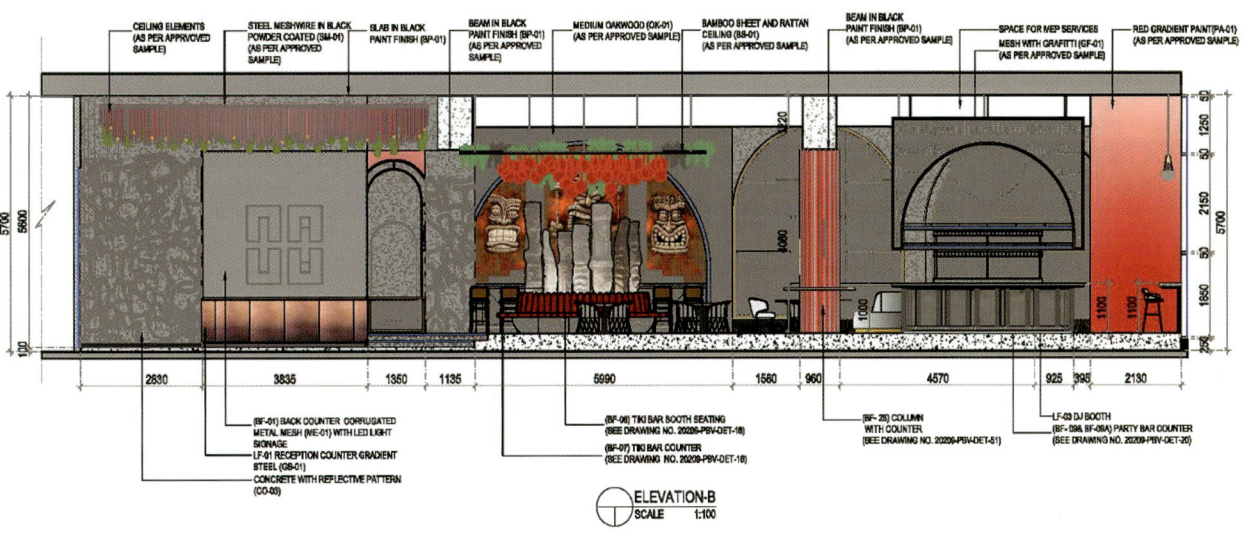

Unique key features - Rum Station
- Tiki bar concept with tiki-style totem poles and art on the walls
- Bamboo surface accents with greenery on the ceiling

Unique key features - Gin point and Vodka Lane
- Retro bar that features an impressive disco ball centered in the arched ceiling
- Black and mild steel colors throughout with black and white terrazzo highlights
- UV paint graffiti on mesh screens

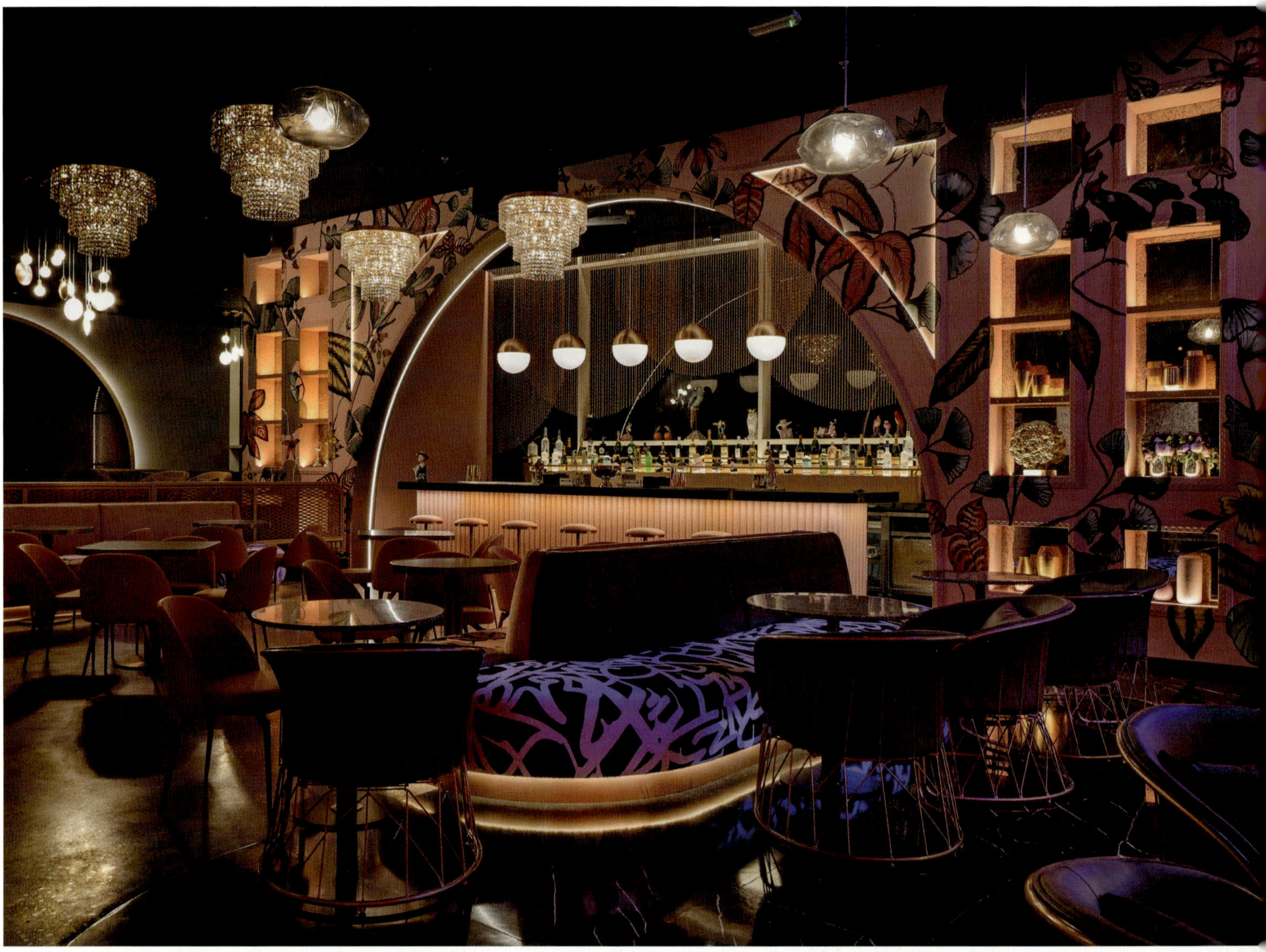

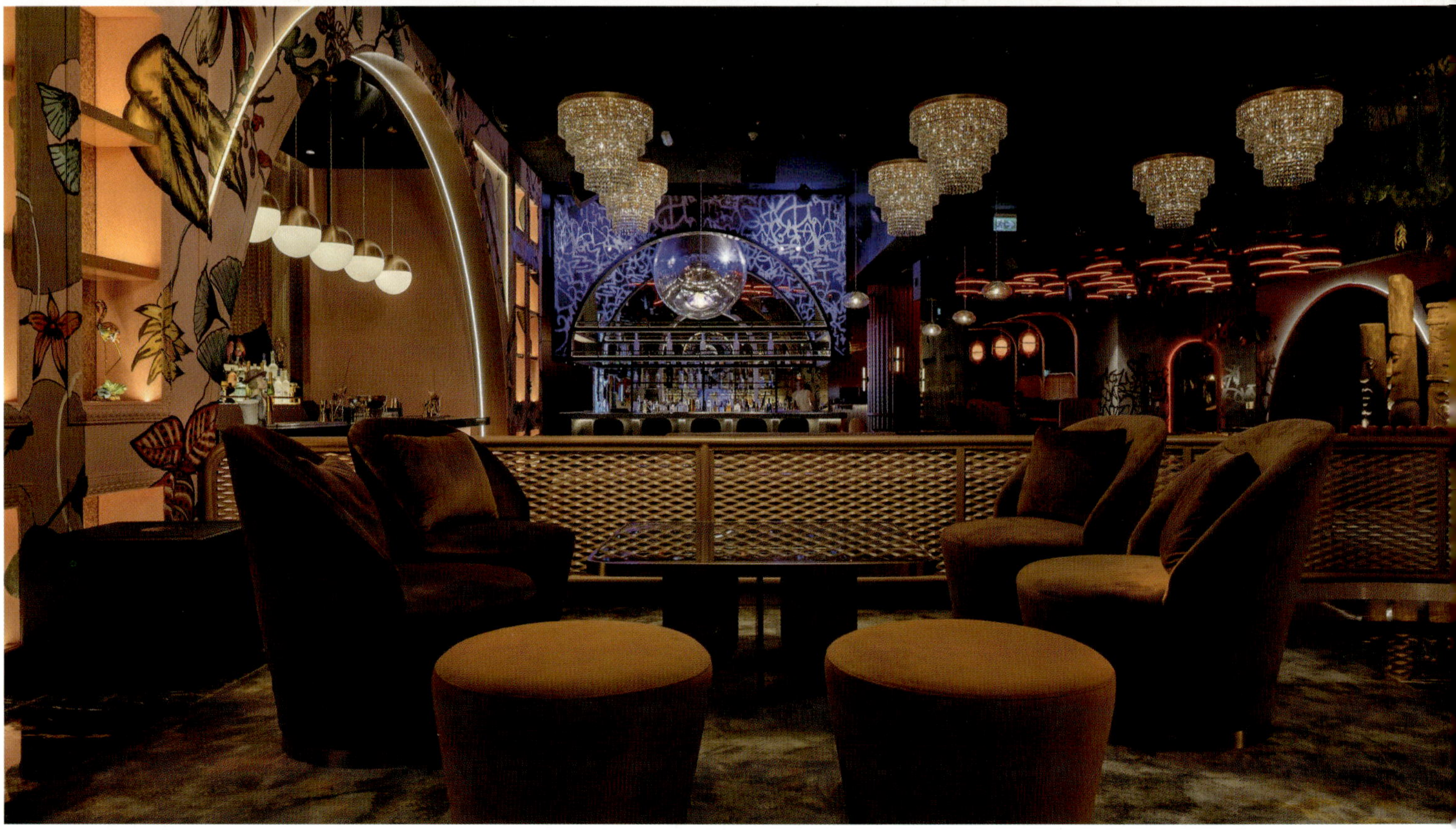

Champagne Avenue
- Pink-hued bar with pink terrazzo highlights
- Bubbly texture with organic patterns used on the arches and moldings and printed on the banquette seating fabrics
- Four classic chandeliers

Unique key features - Sake Alley
- Japanese sake bar
- The wooden structures over the bar are designed and built based on the Japanese Kumiki joinery method
- The bar fascia is made with bricks, stabbed for texture, and filled with plaster
- Careful lighting was chosen to represent a peaceful zen garden. The lighting reflects circular patterns on the floor
- Graffiti on the wall and artworks

Unique key features - Whiskey Square
- The Whiskey Square feels secluded from the other bars and has an imposing arched entry
- A classically designed bar is central in the symmetrical space with sensual onyx lighting
- Inspired by Art Deco style, including lighting and detailing on the fascia of the bar

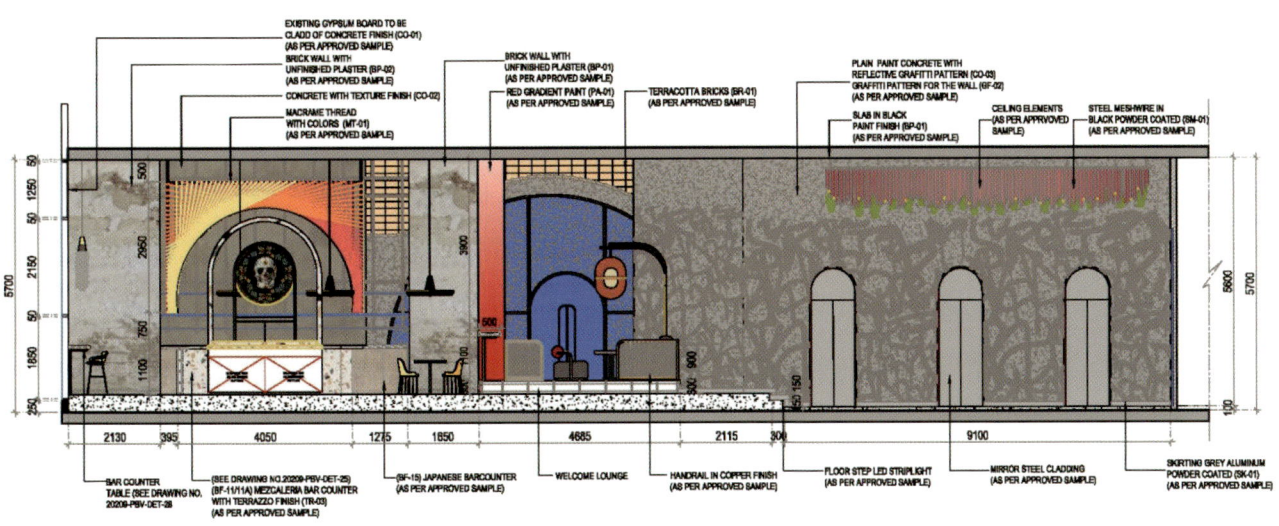

Tequila Road and Mezcal Street
- Designed to capture the spirit of South America with sugar skull centered in the archway
- Colorful terrazzo for the counter and graffiti artwork on the walls
- Lighting inspired by Mexican sombreros

Female Toilet
- Designed with banquette seating and comfortable poufs where ladies can take a rest while waiting in the queue
- The exciting ceiling feature is made from a pink ball installation with stretched mannequin legs

Male Toilet
- Inspired by fun house-style mirrors with led lights and a black ball installation, black tiles with red grouting were used for the walls.

Roma, Italy
VyTA Farnese Bar
VyTA Farnese combines the features of the traditional Italian bar with contemporary design.

Overlooking Piazza Farnese, a stone's throw from Campo de' Fiori, one of Rome's historic squares and the focal point of the capital's nightlife, VyTA Farnese, created by interior design studio COLLIDANIELARCHITETTO combines the features of the traditional Italian bar and contemporary design. The project is an eclectic and experimental transformation of a timeless café: Caffè Farnese, a crossroads for illuminated minds and colourful characters, open from breakfast to post-dinner, where the window of the takeaway bar on Via Dei Baullari invites passers-by to uncurl and unwind.

Daniela Colli has created a space oozing with contemporary charm, fusing the refined elegance of the Renaissance period, European modernism and Millennial pink, with particular attention to detail and exquisite craftsmanship, which distinguishes the VyTA brand in its multiple spaces dedicated to hospitality, in Italy and London. Human experience is the main focus of the project. Interaction with customers is dynamic and guests are at the heart of the space, becoming an integral part of the scene, whether they're perched at the cafeteria, consuming finger-food while sitting on Verpan marsala-coloured leather stools, or comfortably sipping a cocktail on a pink leather sofa.

"VyTA Farnese is an unexpected place, which, with its refined materials and decorations stimulates the interaction between people. In this challenging historical moment, where social distance is the springboard from which to regenerate our lives, this project, through the design of the takeaway bar overlooking the street, and set in one of the world's most beautiful squares, presents a clever solution to hospitality which, as an inescapable expression of sociality, opens up the urban space of our cities", says the architect.

Project details

- **Design:** Collidanielarchitetto
- **Homepage:** https://www.collidaniela.com
- **Area:** 130 m²
- **Location:** Roma, Italy
- **Photography:** Matteo Piazza

About us

COLLIDANIELARCHITETTO is a Rome-based award-winning architecture and interior design studio founded in 2009 by Daniela Colli. Her distinctive hallmark has emerged from a capacity to combine a contemporary vision of society and user needs with extensive knowledge of the historical and cultural roots of interior design, producing results that blend the past with the future.

Her work ranges in scale from furniture to architecture and urban design, with a focus on craft, detailing and precision. She uses the challenges unique to each project, the peculiarities of a site, the specific of a target audience, as catalysts for transformative architecture. As specialists in category – defining hospitality design, her critically acclaimed portfolio spans bars, cafés, restaurants, hotels and spas.

VyTA Farnese was born from the dialogue between place and identity. The project fits into the historical context with refined materials such as polychrome marble, lacquered surfaces, mirrors and velvets, calibrated geometries and symmetries combined with bold shapes and a lively colour palette, from emerald green to Millennial pink with shiny copper accents. The windows on the street frame the interior, revealing a composition of playful textures and metaphysical arches with trompe-l'œil effects. The architecture draws inspiration from the Renaissance period in its use of elementary geometric shapes and for the articulation of the orthogonal and symmetrical plan, while the square and round arch give life to the original elements that characterize the interior design.

VyTA Farnese is developed on two levels: the ground floor houses the bar, the restaurant and the window of the takeaway bar, which allows customers to be served outdoors with piece of mind, and without having to access the premises. Three adjoining spans, connected to each other by large round arches covered in rose gold copper, create a delicate perspective game with increased depth. A bar counter with a central island shapes the matrix of the project: it is the hinge between the two rooms and embodies its intimate atmosphere, emerald green wallpaper and lacquered ceilings. In the corner room, a large arched opening frames the terrace facing Piazza Farnese and its fountain, creating continuity between inside and outside.

Floor Plan

A heavily textured staircase in Alpi green marble, embellished with rose gold copper sphere appliques, leads to the lower level which houses the kitchen and utility rooms. The macroscale square generates the pattern of the hyper-realistic polychrome marble floor. In microscale, it becomes the evanescent decoration of the bar counter, where the overlap of pink mirror, transparent glass and sandblasted serigraphy creates a layer that dissolves, and at the same time amplifies its presence in space.

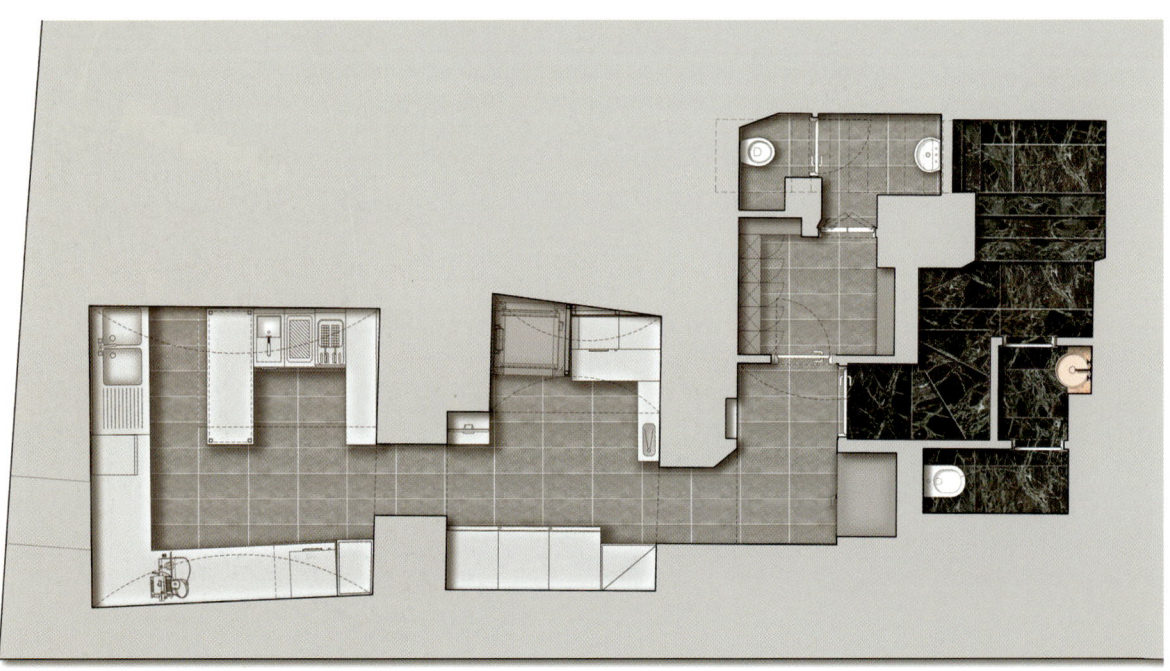

Circular mirrors fragment the reflected images through inserts in white marble, while globes in pink metallized glass illuminate small tables in emerald green glass, and a triptych of arches showcases the wines of the brand. VyTA Farnese redefines the concept of conviviality, giving the city a space that invites guests to breathe outdoors and that looks to the future of hospitality, without forgetting the great tradition of Italian bars and the ritual of coffee consumed standing at the counter as a welcome parenthesis in our bustling daily routines.

217

▶Ostannya Barykada Lviv: площа Ринок, 30, Lviv, Ukraine / +380 73 090 8998

Lviv, Ukraine
Ostannya Barykada Lviv
A gastronomic and artistic space in the Bar & Restaurant

ОБ

The Ostannya Barykada Lviv is a gastronomic and artistic space in the old part of the city, which we worked on for almost a year. This is already the second project under this brand. The first of them has been a visiting card of Kyiv for eight years.

The idea of the OB was born during the Ukrainians' defense of their identity. Therefore, the motif of the interiors is dedicated to the struggle for our freedom and independence. The interior was born from a passionate combination of elements of Ukrainian design identity with signs, materials, and legends against a robust historical environment.

Project details

- **Design:** loft buro
- **Homepage:** https://loftburo.com/
- **Area:** 264 m²
- **Location:** Lviv, Ukraine
- **Photography:** Mykola Korsun

About us

Loft buro, est. 2001, is a creative team, which consists of professional architects, designers and painters.

During this period of time, many projects in interior design and architecture were made by the team. Our main task is the creation of a harmonious, comfortable and cosy space that displays the inner world of a person and creates the perfect mood for all to enter it.

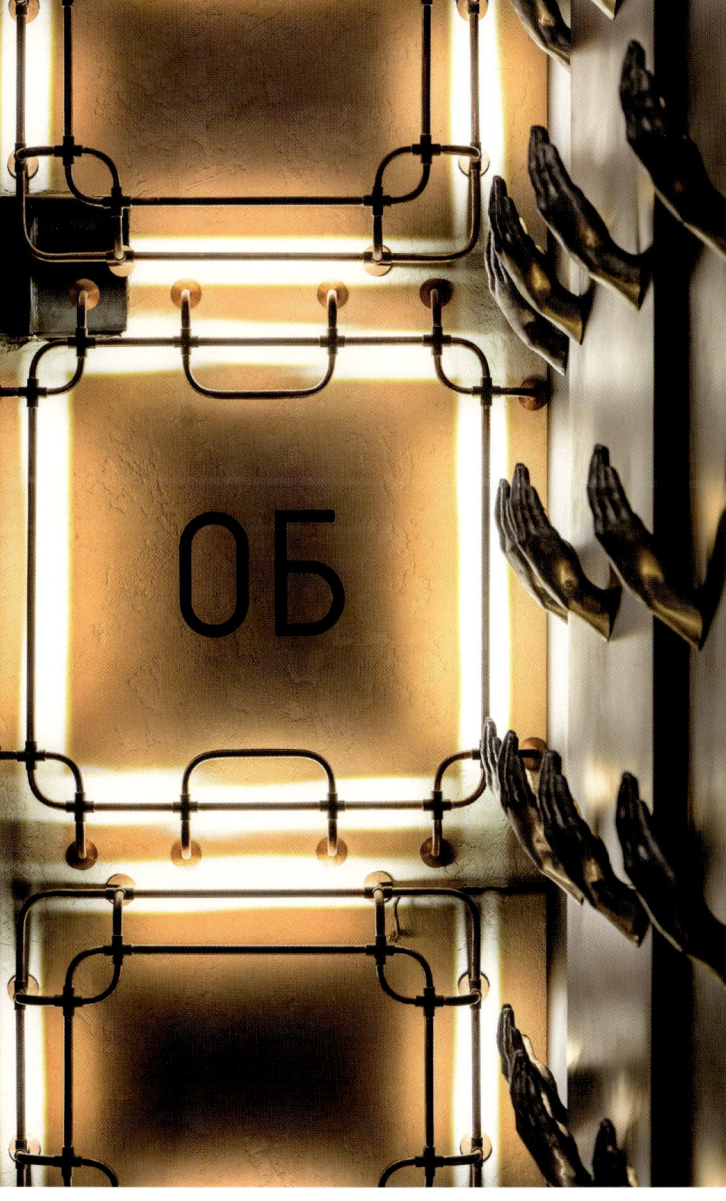
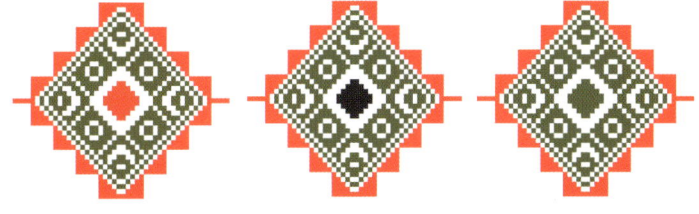

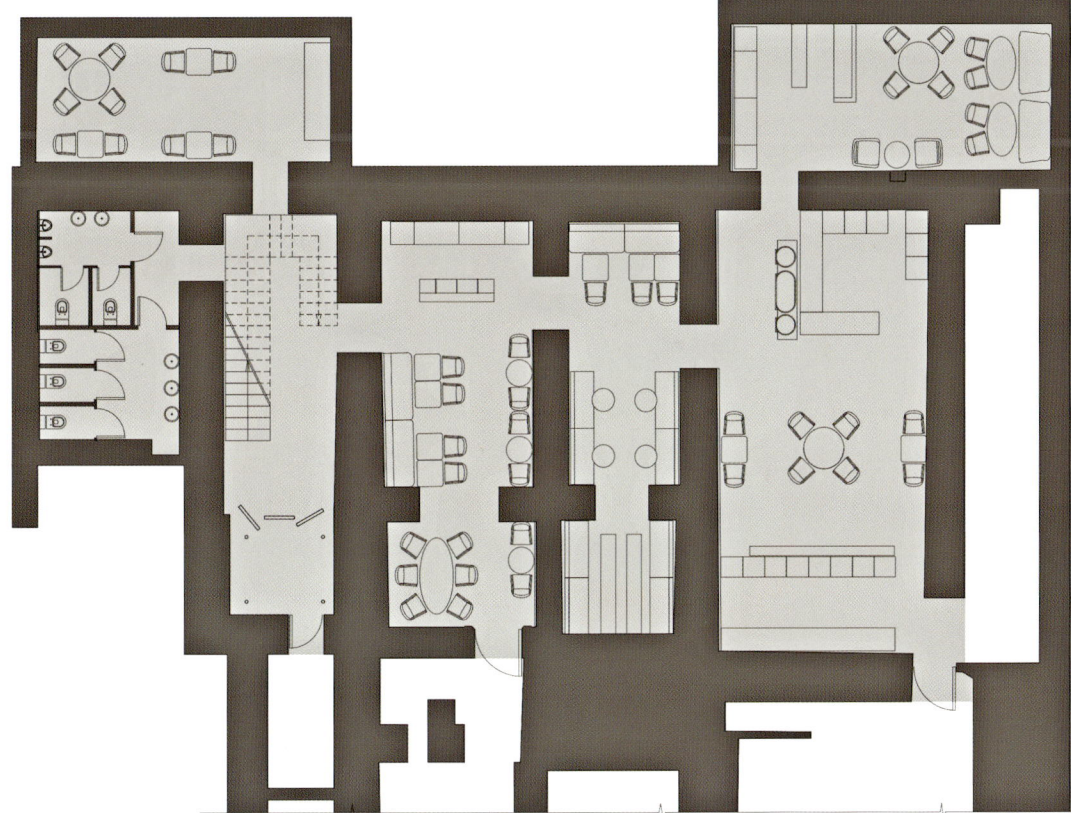

Floor Plan

The idea of the OB was born during the Ukrainians' defense of their identity. Therefore, the motif of the interiors is dedicated to the struggle for our freedom and independence.

Zweig bar
Lviv, Ukraine

One of the most prominent features of the space is the original brickwork groin-vaulted ceiling.

ZWEIG

Zweig bar has recently opened its doors in revitalized premises of the old military plant Arsenal, just a few steps away from the popular Kyiv Food Market and café-bar Molodist. It is always both challenging and rewarding to work on historically-significant buildings. Often structurally complex, such buildings require great care during construction, and warrant a range of limitations as to extent of intended changes.

CONCEPT Zweig is a 210-sq.m space with a 60-guest capacity; you can't but make a connection between the bar's target guests – people with shared interests in art and fine dining, love for travel and reading – and the name of Stefan Zweig, an Austrian novelist, playwright, journalist and biographer, and the most translated author in Europe in the 1920s and 1930s. Leaving the connection question aside, Zweig is sure to attract young intelligentsia and connoisseurs of fine wine and cuisine.

ENGINEERING One of the most prominent features of the space is the original brickwork groin-vaulted ceiling. The decision to keep it clear of any visible utility and communications lines were made on the spot. This brought up a question of how to ensure the ventilation of the premises. We solved this issue by installing vent systems in the flooring, spandrels of the vault, and in-wall recesses.

Project details

▶**Design:** Balbek bureau Architects
▶**Homepage:** https://www.balbek.com
▶**Area:** 210 m²
▶**Location:** Lviv, Ukraine
▶**Photography:** Yevhenii Avramenko

About us

balbek bureau is an award-winning architecture and interior design studio founded by Ukrainian architect Slava Balbek and Borys Dorogov. For 14 years, we have been designing bespoke commercial, corporate and residential spaces.

Comfort, innovation and functionality are the driving forces behind every project we work on. Our approach is to explore the basics and then plunge into details to transform aspirations into ground-breaking environments.

Our work has received multiple international awards and has been published in numerous media outlets worldwide.

BRANDING

Kostia Chikurov and Sasha Blagov from New Agency developed branding for Zweig bar, whereas drawings are from a Kyiv-based illustrator Sergiy Maidukov.

DESIGN A large bookstand stylized as an arch welcomes Zweig's guests and sets up the tone of the place. Its shape bears unmistakable reference to historical architecture, while structural details are refreshingly modern. Three tall openings in the arch lead to the main seating area.

The arch is multi-functional: it serves as a bookcase where you can grab a book to read at your table; it is a retail station where you can buy a book; it is also a partition that separates the entrance zone featuring a soft seating area by the window from the main hall.

DESIGN The focal point of the main hall is the bar with its liquor bottle shelving unit. It is shaped like a wide arch trimmed with sand-colored brickwork. The brick 'frame', colorful bottles on the shelves, and the unit's illumination create a resemblance to a stained-glass window that fits effortlessly in the interior. There are small tables on both sides of the bar, and a large round table in the middle of the hall.

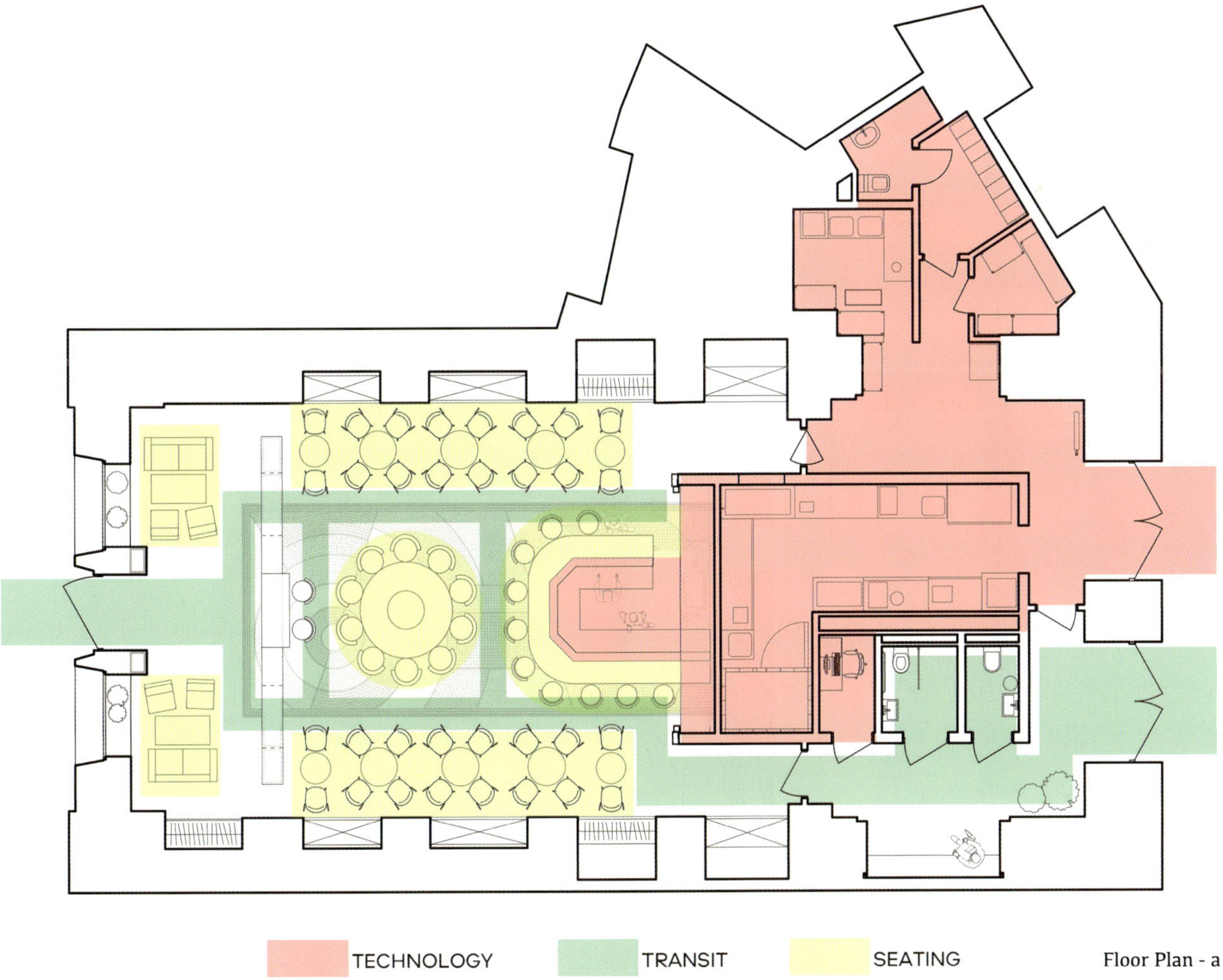

TECHNOLOGY   TRANSIT   SEATING   Floor Plan - a

DESIGN

To preserve the texture of the brickwork while avoiding a typical 'pub interior' look, we applied a thin coat of plaster in a satin finish. This accentuated the complex geometry and texture of the vault and created a warm glow of reflective surfaces, filling up the room with ambient light. Hidden behind a partition of the main hall are guest lavatories and a service zone.

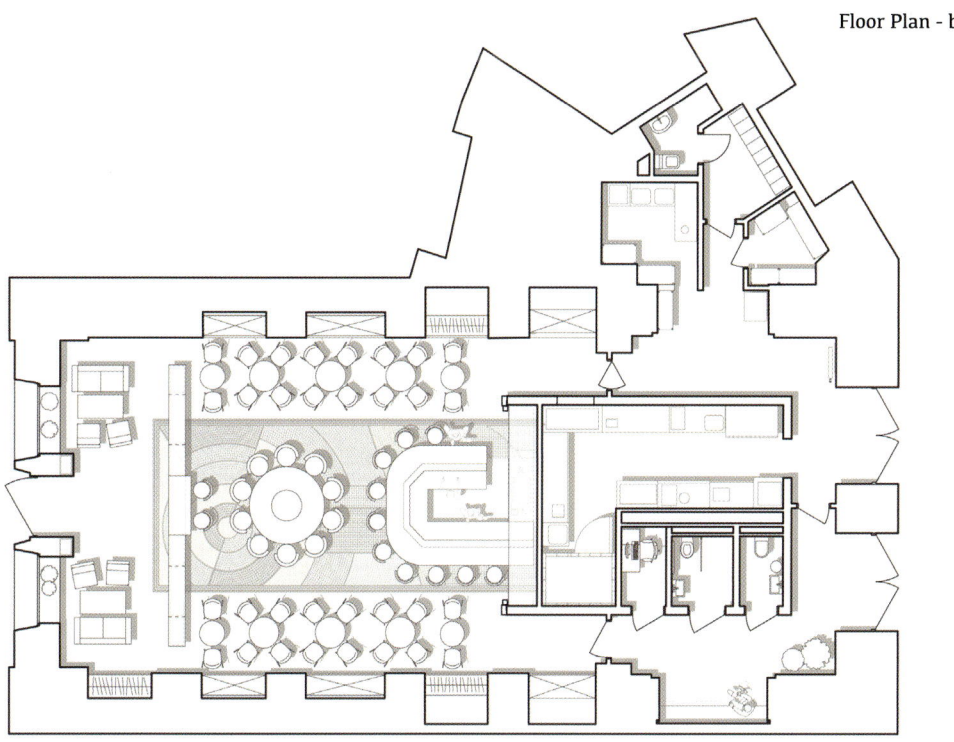

Floor Plan - b

DESIGN There is an atmosphere of harmony and comfort at Zweig that encourages pleasant pastimes over quiet conversations or tête-à-tête with a book. The colors are muted, the light is soft, and the lines are curved. Even hard surfaces – the vaulted ceiling, mosaic floor, the arched bar – are softened with warm shades of grey and honey.

V1

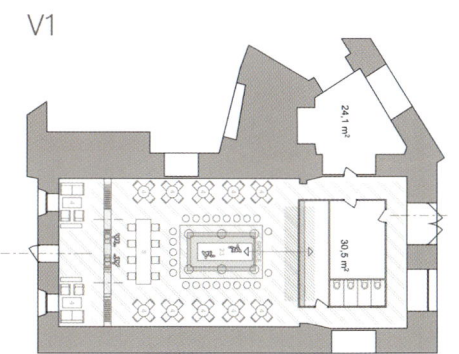

V2

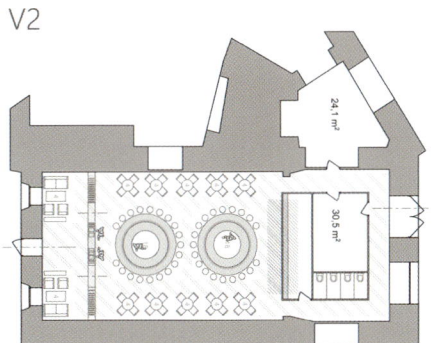

V3

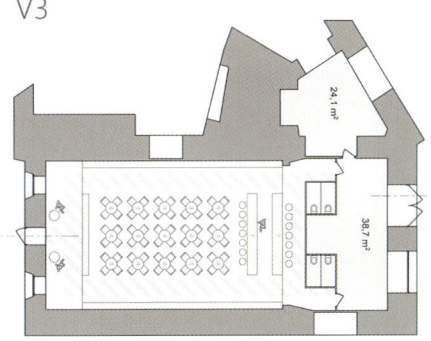

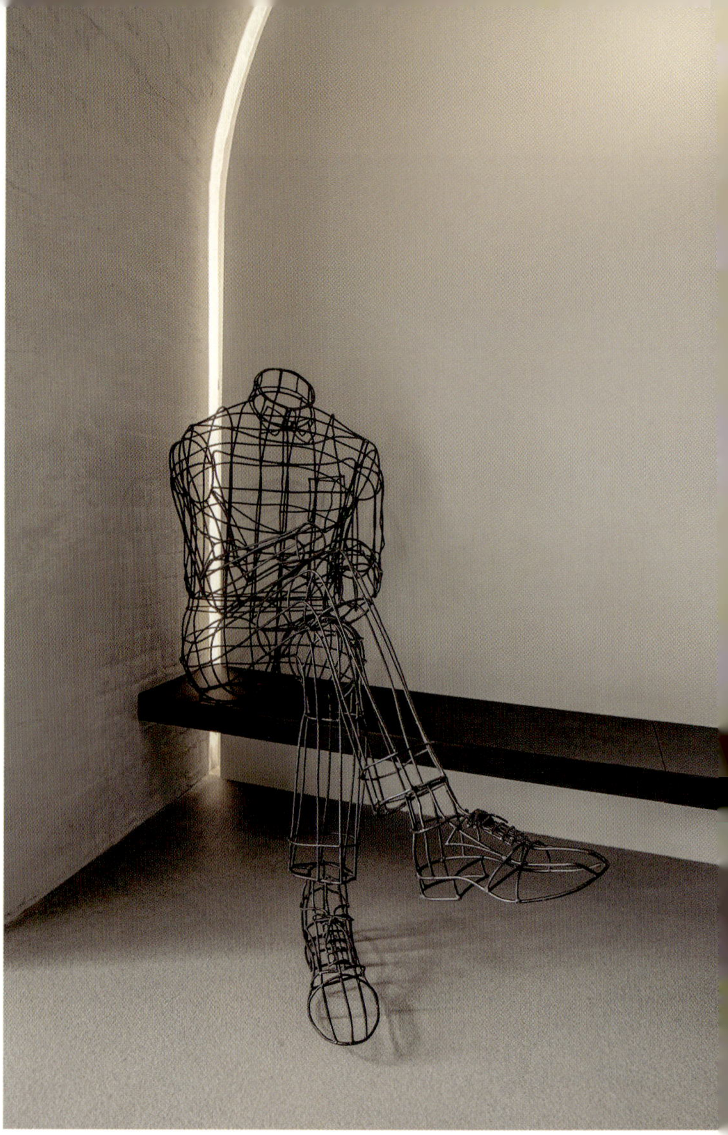

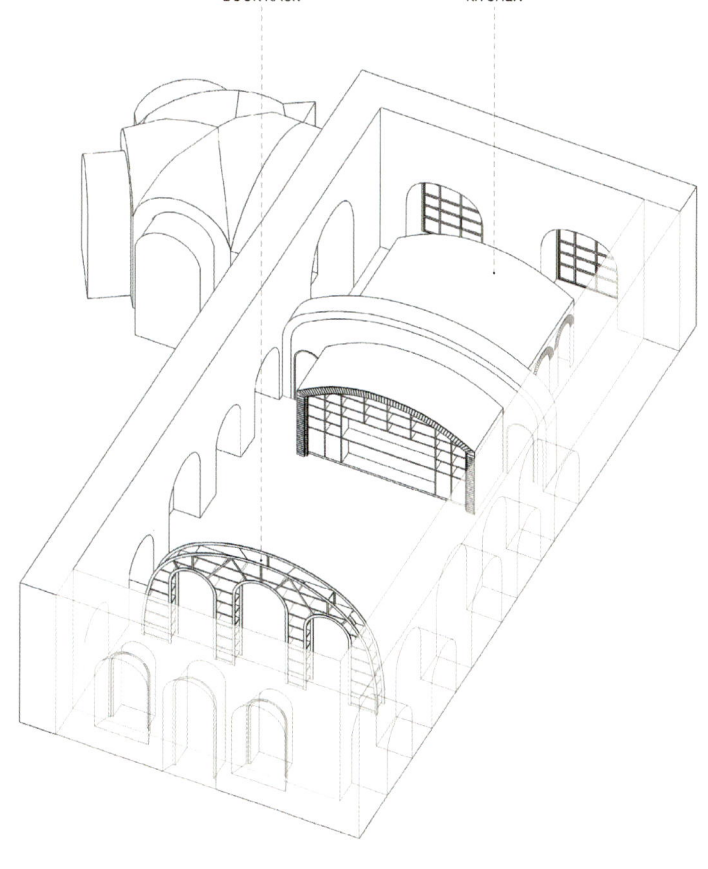

BOOK RACK   KITCHEN

▶Merci Marcel Orchard Road: Уn. 56 Eng Hoon Street, Singapore, Singapore / +65 6224 0113

Orchard Road, Singapore
Merci Marcel Bar
To incorporate a multi-brand concept store into the space

Design Statement: A contemporary and whimsical Paris-style café and bar with a nod to the tropics through the use of colours, textures, design and lighting details, and vintage and modern furniture.

The Site: Having successfully launched the popular Merci Marcel F&B outlets on Club Street and in Tiong Bahru, the client (Antoine Rouland and Marie-Charlotte Ley Rouland of FROG'YS Pte. Ltd.) was ready to launch its flagship café, bar and multi-brand store in Palais Renaissance, a destination mall with unique brands. The client acquired the space previously occupied by a European luxury watchmaker for Merci Marcel Orchard Road, totalling 2377 sqf with a premium 10-metre frontage with a view of the shopping district. HUI DESIGNS is pleased to once again partner with the client to design and deliver its distinctive F&B vision.

The Façade and Patio: With a street view from its space on level one, the client and I agreed that we needed to include a corner dining patio to the design - one that opens to Orchard Road on one side and offers a side view of the Royal Thai Embassy.
This required that we hack the building's external walls. I designed sweeping arches of metal framed windows, walls and doors as a nod to a style popular with bungalows in 1950s Singapore and reminiscent of Bauhaus architecture. This immediately raised the visibility of Merci Marcel Orchard Road and connected the shopping district with the new flagship outlet.

Project details

- **Design:** Hui Designs
- **Homepage:** https://huidesigns.squarespace.com
- **Area:** 198 m²
- **Location:** Orchard Road, Singapore
- **Photography:** Tawan Conchonnet

About us

Siew Hui Lim is the brainchild behind Hui Designs and has over 18 years of International interior design experience. Hui has worked on a wide range of projects including restaurants, bar, high end residential and hotel spaces.

Contacts e-mail: hui@huidesigns.com / +(65) 96371304

■ **The Style:** The style is a contemporary interpretation of "France In The Tropics" with an eclectic mix of mid-century modern and vintage furniture pieces from Europe juxtaposed with 1950s Singapore design and the use of rattan, which was common in warmer climes.

The Café and Bar: The client requested for a foyer at the entrance to create a sense of arrival before customers enter the café and bar. It also functions as a side entrance to the multi-brand store carrying predominantly French products. I designed a welcoming and intimate space with touches of South of France and Morocco. A vintage velvet sofa from Paris invites waiting customers to sit and take in the vaulted ceiling and walls finished in basil green plaster rendering before they are led to their seats.

s customers step into the main space, they will immediately see the island
r, their arrival heightened by the overhanging illuminated brass shelves
led with rows of premium spirits and liqueurs and indoor plants cascading
om different levels.

Fluted solid ash wood strips line the fascia of the bar counter, a detail taken from Merci Marcel Tiong Bahru, the group's first bar in Singapore. Like all other Merci Marcel locations, palm trees, fronds and various types of indoor plants frame the space, softening the mix of rough plaster rendered walls and exposed painted brick walls. The palm theme is seen in the large wall mural painted by French artist Tiphaine Sartini, complementing the quirky art pieces and soft furnishings sourced by the client, Marie-Charlotte. Customers will also quickly notice what the client joyfully calls the Rosé Wall near the entrance window. This feature wall of shelves is wrapped in strips made from rattan, a distinctive feature in Merci Marcel outlets. At the client's request, each shelf has been lined with low-voltage flame bulbs that will cast a soft, gentle glow to bottles of French Rosé on display.

The Style: The style is a contemporary interpretation of "France In The Tropics" with an eclectic mix of mid-century modern and vintage furniture pieces from Europe juxtaposed with 1950s Singapore design and the use of rattan, which was common in warmer climes.

Ceiling: The ceilings of the alcove and patio comprise of woven rattan ceiling panels, which were specially made in Indonesia by BYO Living. Perforated acoustic panels have been used elsewhere.

Floor: For a South of France feel, the hexagon terracotta tiles arranged in honeycomb style are from Italy and supplied by Rubik Material Laboratory.

Decorative Lighting: The hanging lamps are Parasol fringe lamps by Honoré Déco, purchased in Paris. Seletti Monkey lamps have been placed around the café to inject a sense of fun, cheekiness and whimsy.

Furniture: The space hosts an eclectic mix of mid-century modern and vintage pieces, including vintage Danish dining chairs and Saarinen tulip tables and chairs, which were all purchased in Paris. To inject a tropical vibe, rattan-framed sofas reminiscent of 1950s Singapore have been placed around the cafe. Customers will sit on Cesca stools or banquettes that are backless but adorned with cushions from Paris. Like other Merci Marcel venues, the bar will feature Sika bar stools.

Client's brief:

- To design and build Merci Marcel's flagship café and bar on Orchard Road that target "urban bohemians", its third F&B outlet in the country.
- To bring the Merci Marcel brand of contemporary Parisian-style café and bar with a patio dining space that opens to the shopping district.
- To incorporate a multi-brand concept store into the space.
- To incorporate some of the fun and quirky art and objects the clients have collected over the years into the space.

This Merci Marcel ourlet completes that picture – coffee, desserts, wine, food, place to Instagram and shop, and kids would probably feel comfortable here as well. It has this tropical-meets-bohemian chic.

ZWIN & SHOco: Kulparkivska St, 200a, Lviv, Lviv Oblast, Lviv, Ukraine / +380 95 173 3077

Lviv, Ukraine
ZWIN & SHOco

Which combines a modern confectionery-bakery and wine bar

260

"ZWIN" & "SHOco." it's a new multi-format restaurant project in Lviv, which combines a modern confectionery-bakery and wine bar. The uniqueness of this project lies in the combination of two different brands within the same space, which work with different products independently of each other. In the morning operates bakery-confectionery "SHOco." with breakfasts, desserts, pastries, coffee. This is the ideal format for those who have come to work or use co-working space.

In the afternoon there is a transition from the daily café format to the "ZWIN" wine bar, that is a modern space with a focus on the good wine, which aims to develop a wine culture in Lviv. This is a new format space, the mission of which is to show that wine is not just a drink, it's a whole art. It hosts various events related to wine: master classes, meetings with winemakers, tasting for visitors and professionals.

In the design of project used authentic materials: all wooden furniture and installations are made of oak wine barrels. Bar counter and high table tops are lined with copper. Above the bar counter is installed a large-scale installation of 32 Bordeaux barrels with the ZWIN logo.

Project details

▶ **Design:** YOD Group
▶ **Homepage:** https://yoddesign.com.ua
▶ **Area:** 295 m²
▶ **Location:** Lviv, Ukraine
▶ **Photographs:** Andriy Bezuglov
▶ **Graphic design:** Pravda design

About us

Our story began in 2004, when studio's founder Volodymyr Nepyivoda and art director Dmytro Bonesko united their efforts and brought together talented architects and designers. Today we work predominantly in the sphere of commercial design and create projects in hospitality sector: hotels, restaurants, cafes and bars.

Taking on any new project, we want to become a reliable partner who is familiar with restaurant business and shares Client's vision. First of all, we search the idea which will be borne in people's mind and bring long-standing customers to you. To create a perfect image of the object, we design individually furniture, lighting and decoration, implement experimental solutions, create naming, graphic design and dishes serving.

Today we have the best restaurants in our portfolio and successful restaurateurs and hoteliers among our Clients. Our projects are published in famous magazines and win prestigious competitions. But the most important thing is that they bring our Clients love and loyalty of the guests.

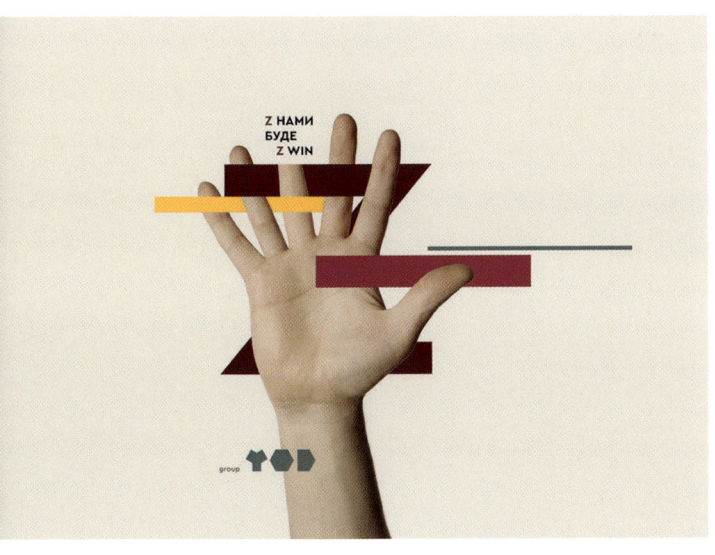

▶ **Contacts** e-mail: yodlab.pm@gmail.com / Call us: +380 (98) 524 85 84

In the design of project used authentic materials: all wooden furniture and installations are made of oak wine barrels. Bar counter and high table tops are lined with copper. Above the bar counter is installed a large-scale installation of 32 Bordeaux barrels with the ZWIN logo.

ZWIN WINE ROBATA BAR

ІГРИСТІ ТА ІГРИВІ
Champagne · Cava

70 kinds of red

ЧЕРВОНІ З НОВОГО СВІТУ

БІЛІ ТА ВИШУКАНІ

70 kinds of white

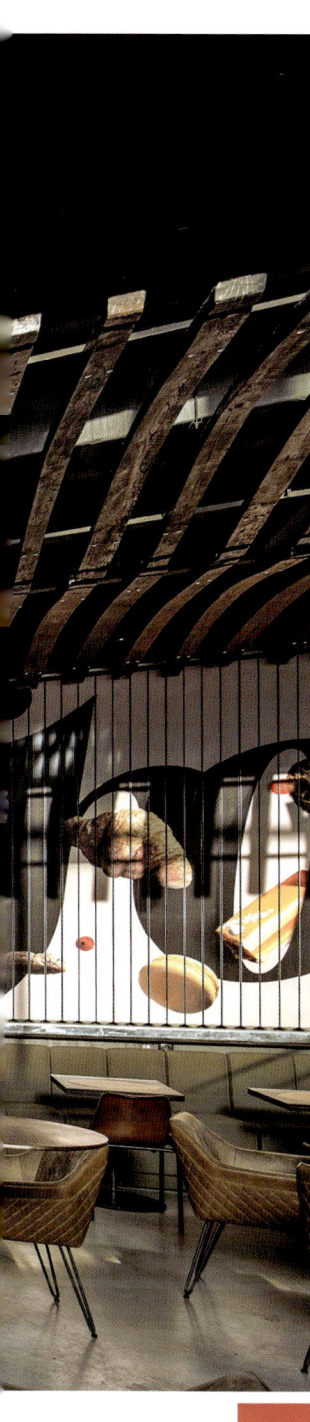

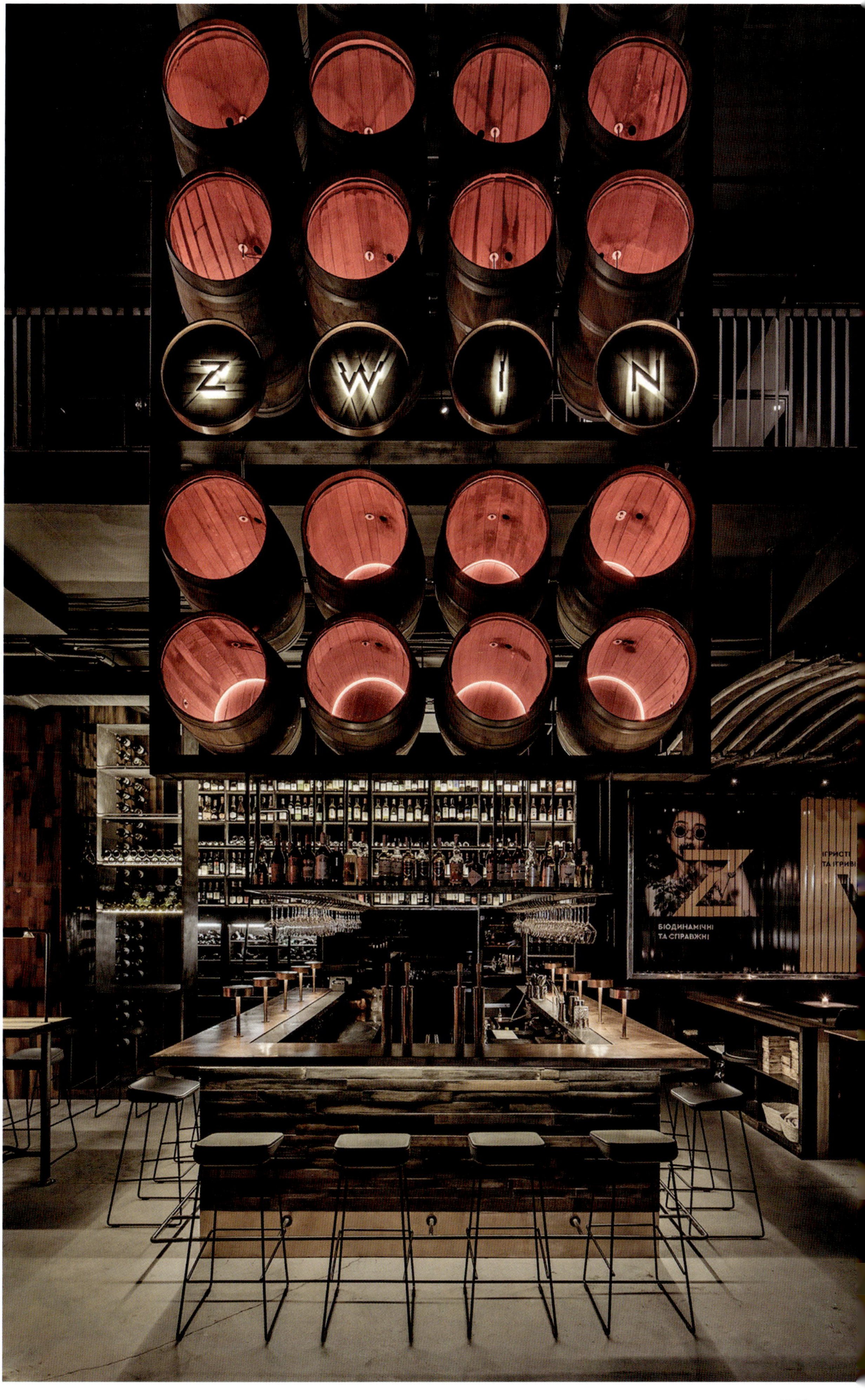
Z W I N
БІОДИНАМІЧНІ
ТА СПРАВЖНІ

Thanks to backlighting, it attracts people's attention from the street – in the evening it is clearly visible through the front panoramic windows. Changing the format in the interior is reflected on the center prismavision. Also, the lighting changes and the active music switches on. Living trees planted in pots emphasize the connection with nature, adding cosiness and comfort to the whole space.

Combining a bakery-confectionery and a wine bar within the same space is a quite successful solution for open space, based on the coworking. «SHOco.» – format for those, who are constantly in business and always in a hurry, who need quick breakfasts and delicious snacks. «ZWIN» provides an opportunity to relax and rest from work without changing location, "slow down" time in the company of selected wine, hearty lunch or dinner and lively music. Thus, both formats harmoniously co-exist, complementing each other throughout the day.

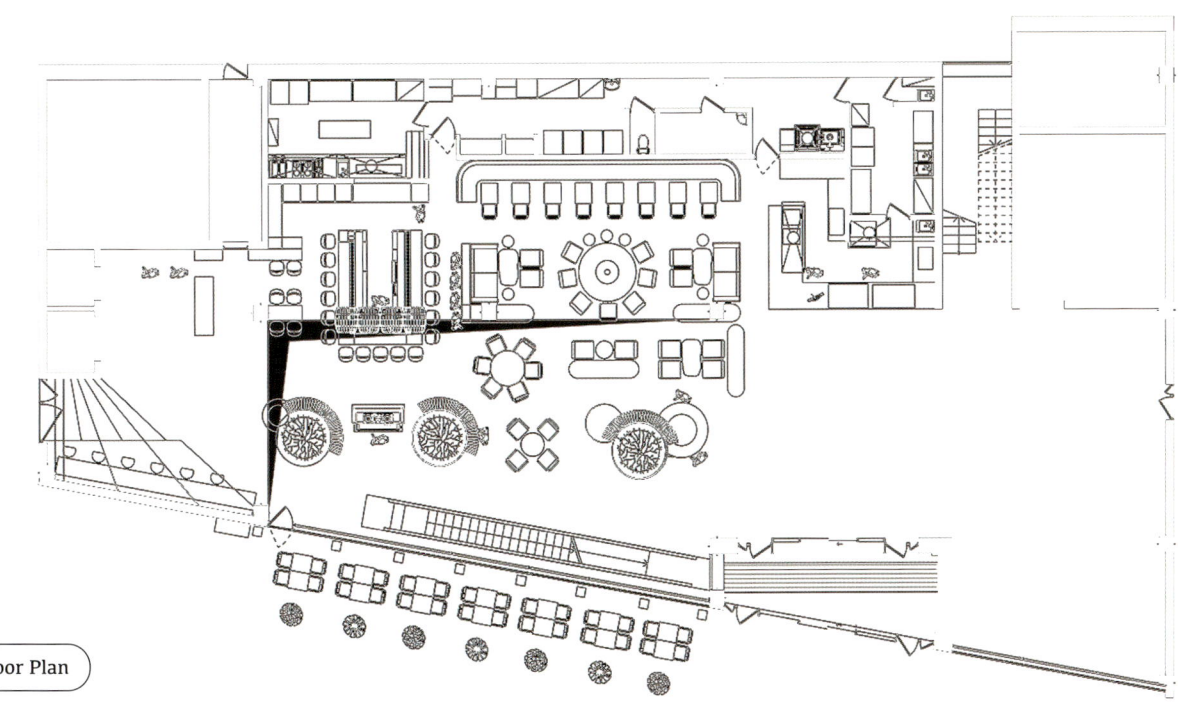

Floor Plan

More Fun Bar
Hangzhou, China

"Rubik's cube" as the design concept. The square and round shapes form an irregular, intriguing polyhedron.

Hangzhou is a renowned leisure city with a vibrant nightlife. The city witnessed a shift from wine shops and tea houses in ancient times to the cocktail culture nowadays that caters to the demands of contemporary young customers. This reflects the evolution of leisure, socializing and life styles of the city. Based in Hangzhou, MORE FUN is a bar conceived by PIG DESIGN led by Li Wenqiang. Drawing inspiration from the "Rubik's cube", the design team reshaped the facade and spatial form and created a series of innovative scenes, to respond to the cultural context and the pattern of streets and alleys in Hangzhou and to create a destination secluded in the old neighborhood.

Mysterious facades:
Located in Jiefang Road adjacent to West Lake, the project occupies parts of the first and second floors of an old residential building at the street corner. The memory of the neighborhood and the fabrics of streets and alleys in its surroundings provide infinite possibilities for breeding future new lifestyles. Inspired by the staggered blocks of the existing building, the designers decided to take "Rubik's cube" as the design concept. Cubic blocks at staggering heights interpenetrate each other, and window openings are subtly carved out on the exteriors. The square and round shapes form an irregular, intriguing polyhedron.

The curtain wall system is formed by thin, prefabricated modular cement panels, rather than cast-in-situ concrete, which realized rapid and low-cost construction within a month. The lightweight modular components also effectively solved the load bearing problems of the old building, and obtained an exterior texture similar to fair-faced concrete. As time goes, the grain on the facades is enriched, and reveals a rough aesthetic complemented by light and shadows.

Project details

- ▶**Design:** PIG Design (pigdesign.art)
- ▶**Homepage:** http://pigdesign.art
- ▶**Area:** 216 square meters
- ▶**Location:** Hangzhou, China
- ▶**Photographs:** SFAP
- ▶**VI design:** Qian Chaofan

About us

Pig Design founded in 2015, the space architecture is established based on the way of literary creation, emphasizing the practical value of art in design application, and opposing the free breeding of traditional aesthetics in space.

Graduated from the Oil painting department of The China Academy of Art, with the urban background as the blueprint, based on the unique perspective and diverse creation forms of artistic creation and architectural interior design, to build interesting space.

Infinite possibilities:

The project is positioned as a fashionable leisure and socializing space serving creative cocktails. Just like the diverse combinations of a Rubik's cube, cocktail making requires subtle coordination among ingredients and the strength and speed of shaking, to present rich sensory experiences through the change of layers and colors.

The interior space of the bar features staggered blocks, which echo the external facades' structures and proportions. Window openings are carved out on the walls based on the arrangement of seating and the needs of introducing light and creating frame views. Set at different heights, those windows create fun visual interaction from different angles.

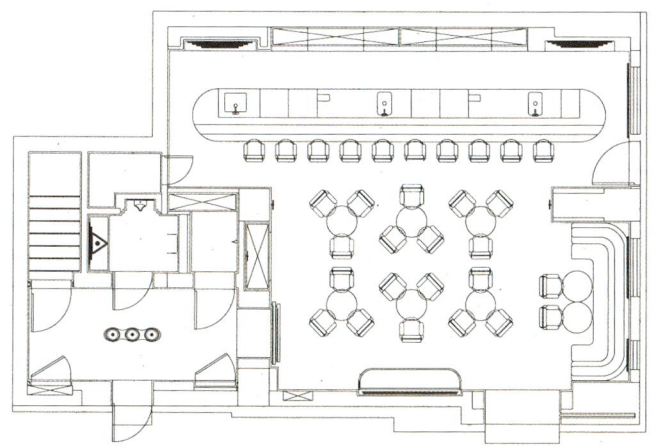

First floor

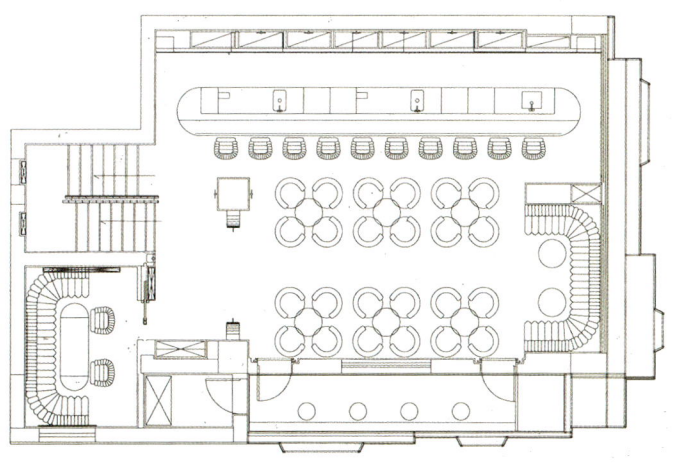

Second floor

The two-floor interior space continues the modular concept of the architectural facades. Taking point, line and plane as basic elements, the designers created various functional geometric forms and decorative surfaces. The space is highlighted by the textural contrast of different materials like brass, washed stones and terrazzo.

With an area of 80 sqm, the first floor presents a neutral tone coordinated by blue, green and gray, and provides an inclusive, leisurely atmosphere suitable for general customers. The linear bar counter, the dotted seats, the unique lighting fixtures at different heights as well as the light strips create diversified sensory experiences, and stimulate customers' desire to explore the space.

The 120-square-meter space on the second floor mainly adopts gray and purple hues, and offers exquisite, mysterious scenes to satisfy the demands of customers with strong brand stickiness for customized services. Circular lines extend from the ceiling lighting to the chairs. Leather and brass generate a retro, industrial tone, and the orderly arrays of mirrored round holes at vertical or horizontal levels create dramatic visual effects.

Simplified solutions:
It took a year to complete the project. At the preliminary stage of the project, the design team dug into the trends of bar sector from the perspective of the brand More Fun. The spatial layout reserves sufficient possibilities for commercial operation, and meanwhile helps release maximum potential to fulfill diversified consumption demands.

The logic of Rubik's cube is a process of changing disorder into order, which is based on its underlying principles to show visible formal variations. Through the process of rotating and recovering a puzzle cube, its various sides are presented.

More Fun is a dynamic space that fuses diversified elements, clarity and vagueness, hardness and warmth. It records daily life, absorbs urban spirit reflected by socializing culture, and writes a chapter about the future unknown world with different people coming here.

Hidden surprises:
The staircase leads to casual "stages" where cool elements and restrained expressions are interwoven. With chamfered corners, the seemingly cold materials show a warm touch. Washed stone veneers with dense grains produce an increasing sense of textures over time. Made of stones and brass, the three cylindrical lighting installations simulate enlarged "candlesticks", and brighten six ancillary functional spaces around. The various visual signs on the doors are like magical keys, featuring downy patterns outlined by straws and rice papers.

Washrooms are arranged in the mezzanine between 1F and 2F. The unified formal languages of the main spaces are extended to this area. In a tranquil enclosed background complemented by indirect lighting, the glossy metal and mirrors create visual order and continuous perceptual experiences, and also generate a sense of warmth and a geometric aesthetic of abstract lines. Mirrors above the washbasin and the reflections in the middle form a cute "cartoon pig". This is a funny surprise that PIG DESIGN conceived for users.

Melbourne, Australia
Pearl, Chablis & Oysters

Old world meets new. Pearl, Chablis & Oysters - Melbourne CBD

▶More Fun Bar: Bourke Streets Mid City shopping centre, Melbourne, Australia / +61 420 783 719

Project Overview

Pearl was born from the themes of 'Allure and Transcendence' with a desire to create a speakeasy style oasis in the heart of the city. With an end-to-end solution encompassing the strategy, branding and interiors - the team were able to conceptualise all aspects of the personality and feel of the venue while celebrating the niche offering of Oysters and Chablis.

Project Brief

Inspired by the deep greens of the ocean and white luminescence of the cinque terre, our client hoped for a space that really placed the premium produce as the hero. We were tasked to create an intimate and luxurious setting that included a custom oyster display and wine fridge, whilst also creating a portal to the neighbouring tenancy - Pinchy's.
The result challenges the idea of what a speakeasy bar should look like. Walking into an environment shrouded from the outside world, it is bright and lofty without compromising on mood and warmth. Dry, mineral and light bodied. Pearl is an environment that does not only showcase its offer but fully celebrates it. Materiality and detail that exudes luminescence, minerality and dry loftiness transcends the patrons and encapsulates the essence of Chablis and Oysters.

Project Innovation/Need

The oyster display directly on the bar top was custom designed to be the focal point. Drainage has been worked into the stone top to make it fully integrated and fit for purpose. Both entry doors have been custom designed, one is constructed from timber that protects the view out to the garish shopping centre lighting. A custom door pull cut from the same stone as the benchtop hints at the environment within. The other door leading to the neighbouring tenancy is a pocket sliding door made from fluted glass that shrouds the pink neon from within Pinchys, a Lobster and Champagne bar adjacent to Pearl. Connecting these two adds intrigue and cross pollination of patrons between the two tenancies.

Project details

▶**Design:** BrandWorks
▶**Homepage:** http://www.brandworks.co
▶**Area:** 45 square meters
▶**Location:** Melbourne, Australia
▶**Photographs:** Alex Reinders @alex_reinders

About us

We're a design circle of strategic creatives in the business of end-to-end design solutions for a global business community that sees its brands advancing the quality of tomorrow. We work with business leaders and entrepreneurs who believe design can bring transformation and impact within Hotel & Hospitality, Lifestyle, Health & Wellness, and Technology.

Pearl

The Brief

Our client Jeremy Schick (of Pinchy's lobster and champagne bar just next door) dreamt of a concept wine bar that championed another of his favourite pairings in the heart of Melbourne. The desire was for it to be bright, airy but exude warmth and sophistication, just like the produce. "With such a unique and niche F&B concept, we knew we wanted to create more than just a wine bar. Pearl was to be a space that fully celebrated the symbiosis between Chablis and Oysters. The interior palette for Pearl therefore honours the characteristics of her premium offering that you experience on the tasting palette"

Design Challenge

The tight footprint was a significant challenge as we had an ambitious brief for only 45 square metres. This was tackled by placing the patrons up against the wine store and making the doors operable above bar height. Seats have also been placed along the bar where patrons can enjoy oysters being shucked, prepared and presented right in front of them.

Pearl

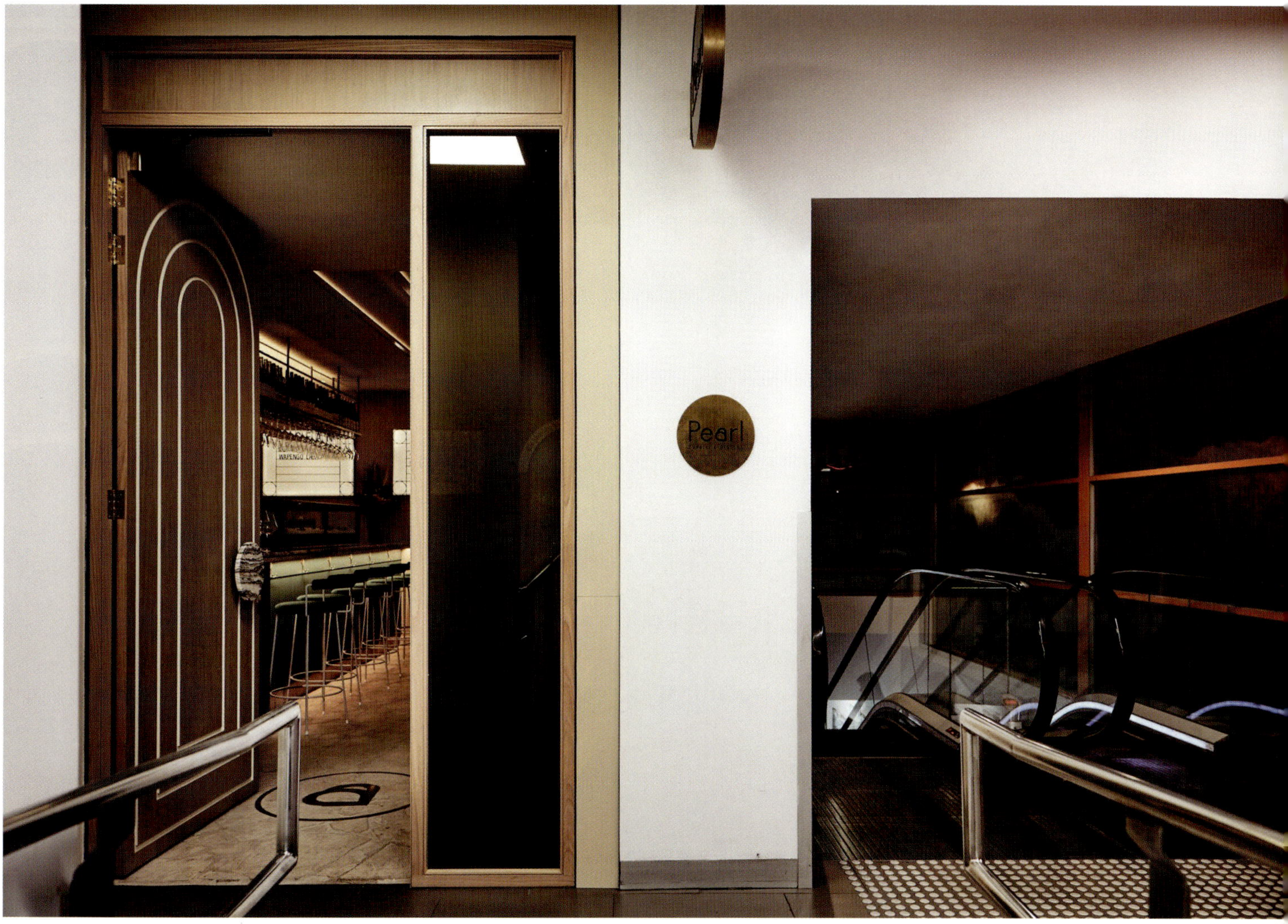

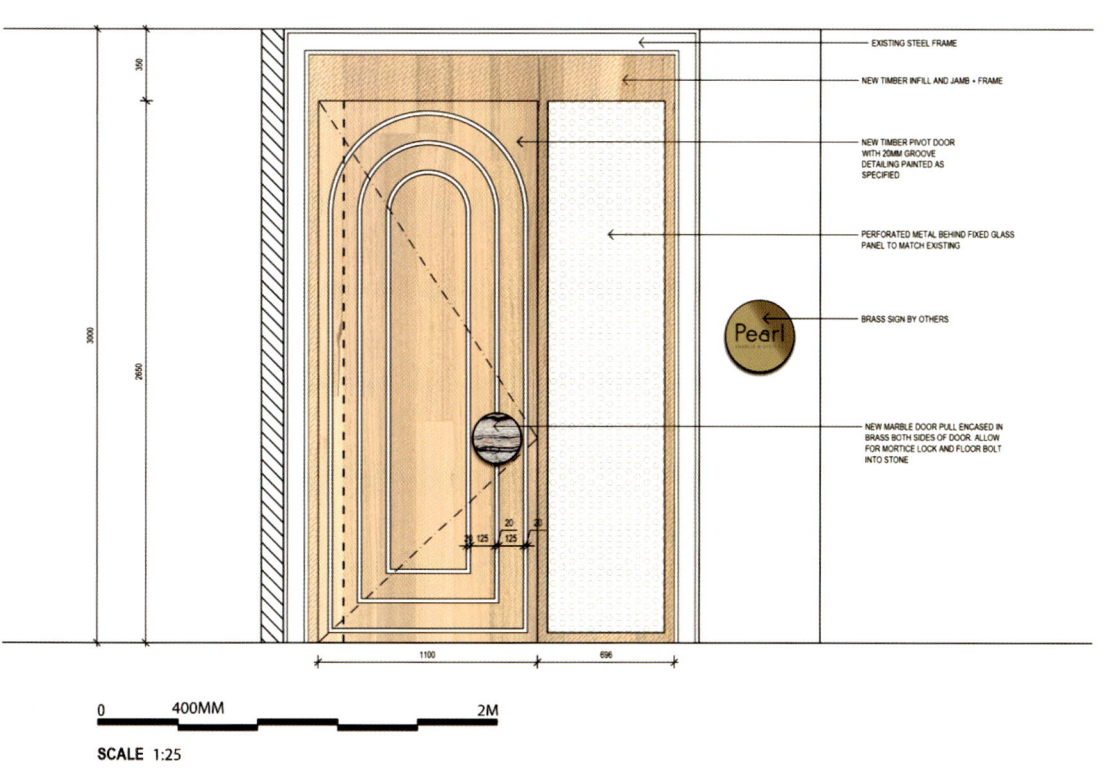

EXISTING STEEL FRAME
NEW TIMBER INFILL AND JAMB + FRAME
NEW TIMBER PIVOT DOOR WITH 20MM GROOVE DETAILING PAINTED AS SPECIFIED
PERFORATED METAL BEHIND FIXED GLASS PANEL TO MATCH EXISTING
BRASS SIGN BY OTHERS
NEW MARBLE DOOR PULL ENCASED IN BRASS BOTH SIDES OF DOOR. ALLOW FOR MORTICE LOCK AND FLOOR BOLT INTO STONE
Pearl
350
3000
2650
20
125
20
125
20
1100
696
0
400MM
2M

SCALE 1:25

Design Challenge

Making a bright and welcoming space that is completely shut off from natural light made for a difficult finishes brief. Following inspiration from outdoor venues and the loftiness of European piazzas we were able to bring the outside in. The travertine crazy paving and Jurassic travertine bar top are strong organic pieces that are complemented by the natural hue of the Victorian Ash. These all work together to give warmth from the natural world in a space completely shut off from it. By including tones of ocean greens this is calmed and complemented.

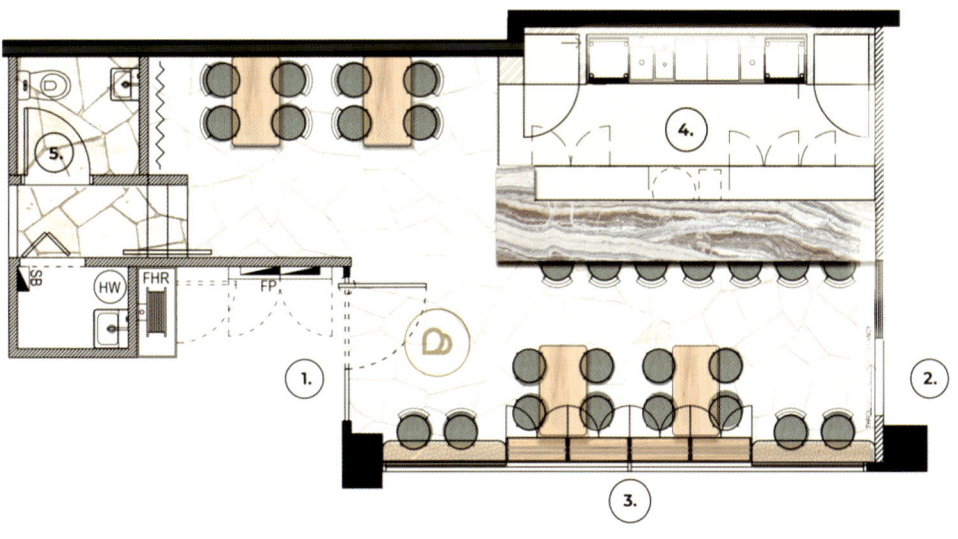

P.01	PROPOSED PLAN

0 — 1 — 5M

SCALE 1:50

KEY

1. MAIN ENTRY
2. NEIGHBOURING BAR ENTRY
3. CHABLIS DISPLAY
4. OYSTER BAR
5. BATHROOM

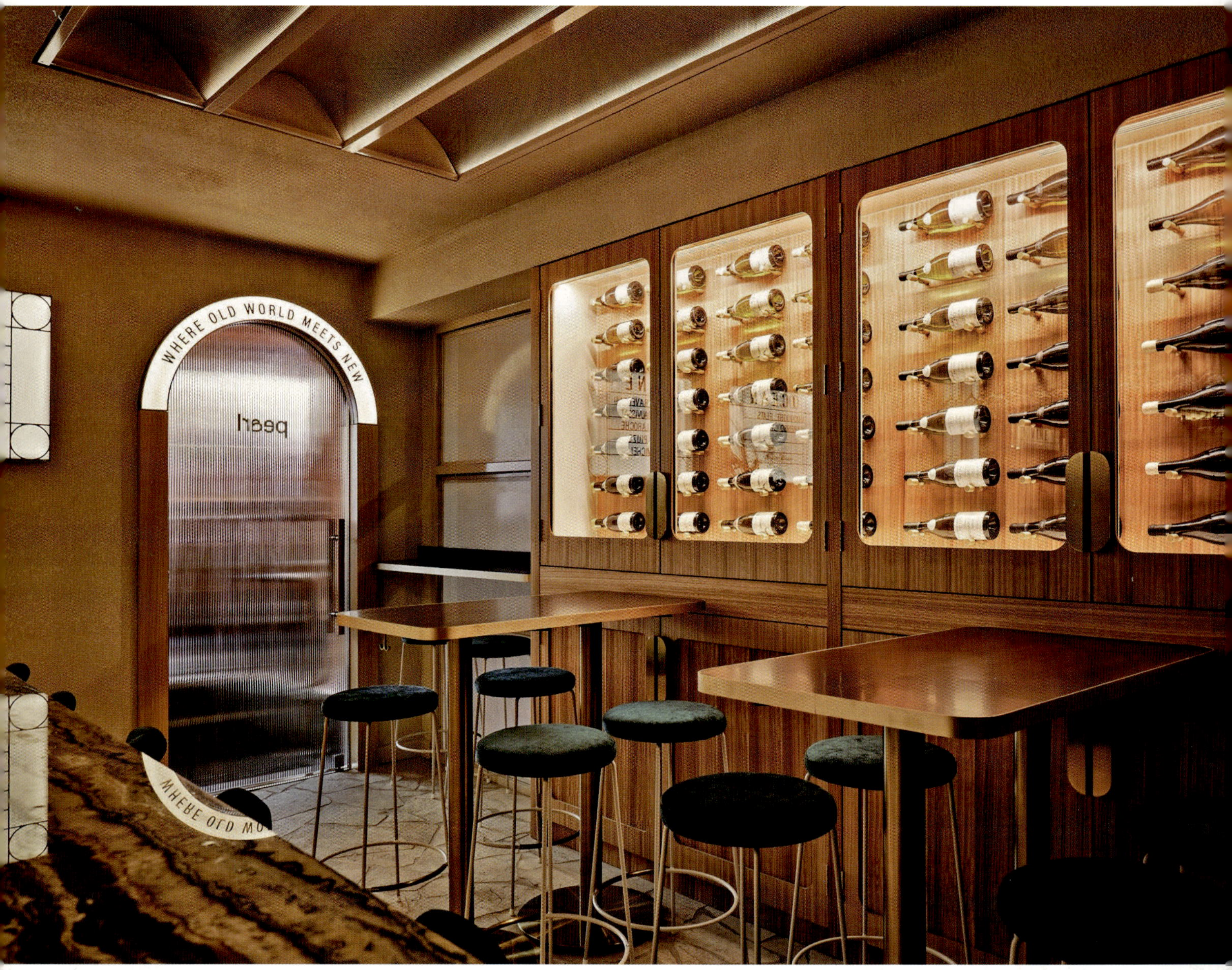
WHERE OLD WORLD MEETS NEW
pearl

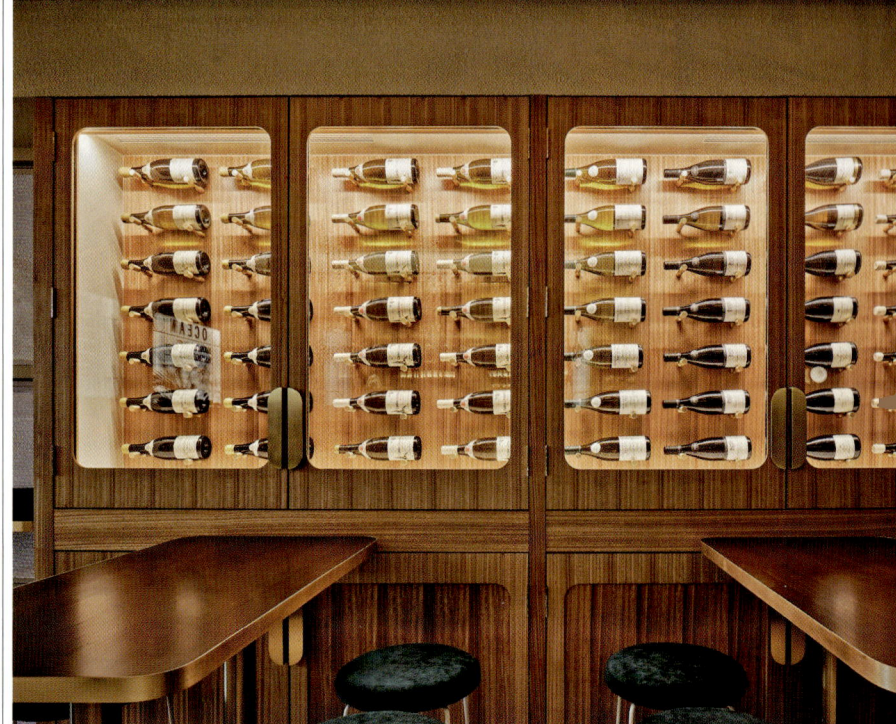

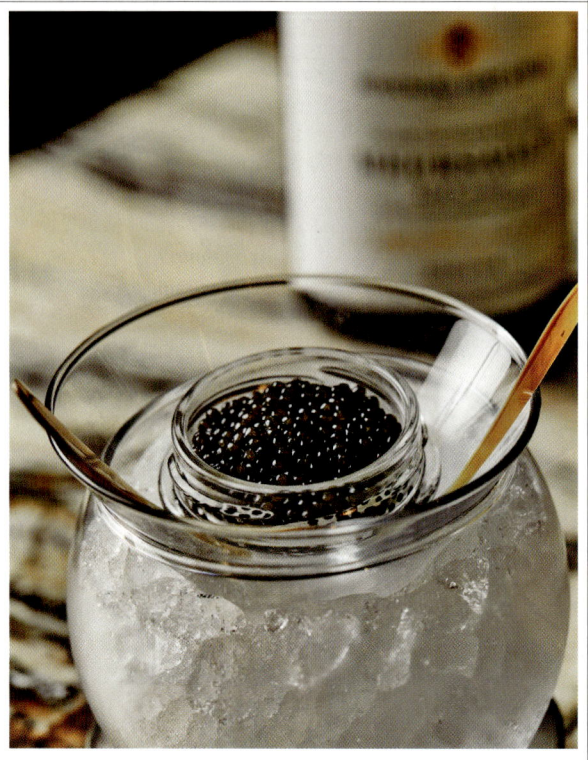

Visible through its large windows on St-Denis street, the focal point of this project is a huge, entirely bespoke, luminous ceiling inspired by art-deco. . This distinctive element bathes the space in a sweeping light and creates a singular atmosphere.

For this project, the APPAREIL Architecture team received the mandate to design an extension for a family home and renovate its shared spaces. The project's initial phase was to develop an extension to the house which acted as a "luminous box" Taking advantage of the lighting, a dining room was outfitted and offers, throughout the year, a unique view on the backyard. Inside, the kitchen was redesigned to foster dialogue between old and new elements. Responding to the owners' wish to preserve the dark tinted floors, the integrated materials are more luminous and neutral. The smooth cupboards with no visible hardware reinforce the space's soberness. In contrast with this immaculate white environment, a central wooden module stands out by its surface textured with wooden slats. Made of walnut, the element brings a touch of warmth and harmonizes with the existing wooden floor. It allows for all essentials to be concealed by acting both as a dish cabinet and pantry. The space is thus unobstructed and seems more spacious.The intervention further extends throughout the ground floor to open the space and create a welcoming relationship between the rooms.

Project details

▶**Design:** APPAREIL architecture

▶**Homepage:** https://www.appareilarchitecture.com

▶**Area:** 2025sqf

▶**Location:** Montreal, QC, Canada

▶**Photographs:** Félix Michaud

About us

APPAREIL Architecture designs an architecture well rooted in our nordicity where each element refers to its essence. A timeless architecture strongly inspired by our origins. A vision of sustainability inhabits each of the projects in order to last in the future. Questioning each of the scales and details of a space is our nature and our methodology, from the envelope to the furniture, from design to completion.

▶Contacts e-mail: info@appareilarchitecture.com / Call us: +438 875-6960

To balance the bar's narrow shape, it was separated into many sub-sections, allowing for 70 seated places to be offered. The space is inhabited by noble materials, infusing character and warmth into the project. Beneath a canvas of black bricks, textured terracotta slats contrast with the smooth brass elements dressing the bar. Touches of ash wood and green marble create an atypical composition, wherein noble materials are skillfully balanced with wood's relaxed nature. In contrast with this, the bathroom transports clients into a completely monochrome universe. Painted entirely in pink, the glossy ceramics harmonise with textured slats that harken back to the bar.